Accreditation of Employee Assistance Programs

Accreditation of Employee Assistance Programs has been co-published simultaneously as *Employee Assistance Quarterly*, Volume 19, Number 1 2003.

Accreditation of Employee Assistance Programs

R. Paul Maiden, PhD
Editor

Accreditation of Employee Assistance Programs has been co-published simultaneously as *Employee Assistance Quarterly*, Volume 19, Number 1 2003.

Routledge
Taylor & Francis Group
NEW YORK AND LONDON

First published by
The Haworth Press, Inc.
10 Alice Street
Binghamton, N Y 13904-1580

This edition published 2011 by Routledge

Routledge
Taylor & Francis Group
711 Third Avenue
New York, NY 10017

Routledge
Taylor & Francis Group
2 Park Square, Milton Park
Abingdon, Oxon OX14 4RN

Accreditation of Employee Assistance Programs has been co-published simultaneously as *Employee Assistance Quarterly*, Volume 19, Number 1 2003.

The development, preparation, and publication of this work has been undertaken with great care. However, the publisher, employees, editors, and agents of The Haworth Press and all imprints of The Haworth Press, Inc., including The Haworth Medical Press® and Pharmaceutical Products Press®, are not responsible for any errors contained herein or for consequences that may ensue from use of materials or information contained in this work. Opinions expressed by the author(s) are not necessarily those of The Haworth Press, Inc. With regard to case studies, identities and circumstances of individuals discussed herein have been changed to protect confidentiality. Any resemblance to actual persons, living or dead, is entirely coincidental.

Cover design by Marylouise E. Doyle

Library of Congress Cataloging-in-Publication Data

Accreditation of employee assistance programs / R. Paul Maiden, editor.
    p. cm.
    Accreditation of Employee Assistance Programs has been copublished simultaneously as Employee assistance quarterly, Volume 19, Number 1 2003.
    Includes bibliographical references and index.
    ISBN 0-7890-2643-0 (hard cover : alk. paper) – ISBN 0-7890-2644-9 (soft cover : alk. paper)
    1. Employee assistance programs–United States–Evaluation. 2. Employee assistance programs–Canada–Evaluation. 3. Employee assistance programs–Standards–United States. 4. Employee assistance programs–Standards–Canada. I. Maiden, R. Paul. II. Employee assistance quarterly.
HF5549.5.E42A27 2004
658.3'82–dc22
                                          2004015698

Accreditation of Employee Assistance Programs

CONTENTS

ABOUT THE EDITOR

R. Paul Maiden, PhD, is Director of the School of Social Work at the University of Central Florida in Orlando. He teaches Social Policy, Macro Practice, Interventions with Substance Abusers, Strategies in Employee Assistance Programs and directs study programs to Russia and South Africa. He has also taught Strategic Change Management in Public Affairs, and Ethics in Public Affairs in the PhD program. Additionally, he developed and coordinated the graduate addictions certificate program and was also graduate coordinator of the MSW program prior to being appointed director. From 1986-1999 he was Chair of the Occupational Social Work Program at the Jane Addams College of Social Work at the University of Illinois–Chicago. As program chair he developed curriculum in substance abuse treatment, employee assistance programs, managed care, and occupational policy and services. Additionally, he had responsibility for developing and serving as faculty liaison for all EAP field placements as well as coordinating a post-masters EAP certificate program. From 1993-1997, Dr. Maiden directed a U.S. Information Agency funded faculty exchange program with the University of Witwatersrand in Johannesburg, South Africa, where he has been a visiting scholar on several occasions. He has also guest lectured at the University of Pretoria, the University of Natal in Durban and the University of Zimbabwe in Harare. He also initiated and directed a masters and doctoral scholarship program for members of the South African Black Social Workers Association. Other faculty positions include the Virginia Commonwealth University and Abilene Christian University (Texas) and Doctoral Fellow at the University of Maryland.

Dr. Maiden has presented papers and conducted training at numerous national and international conferences. He has published extensively in the areas of employee assistance programs, substance abuse, workplace legislation, evaluation of work-based human services, AIDS in the workplace, alcohol abuse and domestic violence and managed behavioral health care. He is the editor and a contributing author of *Global*

Perspectives of Occupational Social Work (2001), *Employee Assistance Services in the New South Africa* (1999), *Total Quality Management in EAP* (1995) and *Employee Assistance Programs in South Africa* (1992). He is currently the editor of the *Employee Assistance Quarterly* and is on the editorial board of the *EAP Digest*.

Dr. Maiden is a principal of Behavioral Health Concepts, Ltd. He has consulted with a wide range of organizations and employers in the public and private sectors in the development, administration and evaluation of employee assistance and managed care programs, workplace policies and educational programs on drug testing, family medical leave, HIV/AIDS, harassment and disabilities, training and development of treatment providers in managed care, and organizational development and change.

For the past 14 years Dr. Maiden has consulted with numerous organizations in Southern Africa on the development of employee assistance programs to address workplace alcohol, drug, and mental health problems, HIV/AIDS and related health care costs. He also coordinates similar efforts between U.S. companies doing business in South Africa. He has led several delegations of health and human service and education professionals to examine the health, education and welfare delivery systems in Southern African countries and developed a health care trade mission to South Africa for the U.S. Department of Commerce's International Trade Administration. He has traveled extensively in South Africa, Namibia, Botswana, Zimbabwe and Swaziland. He is also currently involved in workplace substance abuse and domestic violence projects in Russia and recently received a Senior Fulbright to develop EAP curriculum and field-based training and consultation in the petroleum industry.

Dr. Maiden is a licensed clinical social worker and has been an active member in the National Association of Social Workers for 25 years. He is a long-time member of the Employee Assistance Professionals Association and is the past president of the Illinois Chapter. He is also a member of the Employee Assistance Society of North America. He holds a masters of social work from the University of Tennessee and a PhD from the University of Maryland School of Social Work.

Preface:
Certification, Licensure, and Accreditation
in Employee Assistance Programs

All reputable educational institutions, organizations, and professional practitioner groups are recognized by independent and autonomous accreditation, licensure or certification and therefore for recognition and status through some form of national accreditation. Professional practitioners such as physicians, lawyers, psychologists and social workers are sanctioned to practice while adhering to specified standards and codes of ethics, practice skills and ongoing professional development which is often recognized and acknowledged through state-administered licensure. Certification is another substantial measure of ascertaining specific competencies and skills required to successfully apply knowledge, tasks and responsibilities involved in specific job descriptions and work assignments.

The field of employee assistance programming has evolved over four decades to a point where certification is available through the Employee Assistance Professionals Association (EAPA), the leading trade and membership group representing employee assistance practitioners. Licensure of EAP practitioners is also occurred in a limited number of states. Organizational accreditation of employee assistance programs has also continued to evolve. The first program accreditation was devel-

[Haworth co-indexing entry note]: "Preface: Certification, Licensure, and Accreditation in Employee Assistance Programs." Maiden, R. Paul. Co-published simultaneously in *Employee Assistance Quarterly* (The Haworth Press, Inc.) Vol. 19, No. 1, 2003, pp. xvii-xx; and: *Accreditation of Employee Assistance Programs* (ed: R. Paul Maiden) The Haworth Press, Inc., 2003, pp. xv-xviii. Single or multiple copies of this article are available for a fee from The Haworth Document Delivery Service [1-800-HAWORTH, 9:00 a.m. - 5:00 p.m. (EST). E-mail address: docdelivery@haworthpress.com].

http://www.haworthpress.com/web/EAQ
xv

oped by the Employee Assistance Society of North America (EASNA), a model adopted late by the Council on Accreditation (COA). An alternative EAP program accreditation procedure was also developed by the Commission on Accreditation of Rehabilitation Facilities (CARF) which is conducted in conjunction with social service agencies also offering employee assistance services.

This volume addresses the evolution of practitioner certification and the development of a comprehensive system of employee assistance program accreditation.

The history of EAP certification, licensure and accreditation is presented by Eddie Haaz, John Maynard, Steve Petrica, and Charlie Williams in "Employee Assistance Program Accreditation: History and Outlook." This article examines certification and accreditation in the EAP field in the U.S. and Canada by EAPA and EASNA. They note that the two professional associations, driven by divergent philosophies, have evolved differently in their approach to accreditation. These two organizations share the conviction that control of standards is essential to the self-definition of a professional field, and has implications as well for marketing and governmental regulation. They suggest that accreditation has an important role and should define acceptable standards in the emerging employee assistance environment, which now also includes managed behavioral health care, work life, and international programs.

Stephanie Pacinella, Assistant Director of Standards Development and Performance Measurement, at the Council on Accreditation (COA), in her article "Developing Standards for Accreditation" suggests that standards for development is a continuous process that relies on an inclusive, consensus-building methodology to ensure that standards maintain relevance in an ever-changing field. This article provides an overview of the framework for the COA's EAP standards, and details the steps in the standards development process that resulted in both the first and current editions of the *COA EAP Standards and Self-Study Manual.*

Tim Stockert, EAP Manager at the Council on Accreditation (COA), in his article "The Council on Accreditation Employee Assistance Program Accreditation Process" describes accreditation as a time-limited, facilitative step-by-step process that involves an internal and external review of an organization's policies, procedures, and practices based on standards of best practice. This article provides an overview of the steps in COA's EAP accreditation from application to reaccreditation.

Paula M. Cayley, Ulrike Scheuchl, and Anne Bowen Walker of Interlock Employee and Family Assistance Corporation of Canada present one of two case studies depicting firsthand experiences of preparing for and guiding their organizations through accreditation. They note that the process of accreditation was an extensive, often challenging, but ultimately exhilarating experience. It provided opportunities to grow as a company and led to the development of a number of new and improved systems and practices. The Interlock group attempt to define their strategy that led to achievement of accreditation and offer some useful guidelines for future applicants.

The second accreditation case study presented by Tina Thompson, Vice President of Employee Assistance Programs and Addictions Services at Magellan Behavioral Health, outlines her company's successful accreditation effort. Thompson discusses Magellan's experience and lessons learned while going through such a process.

Dale Masi, Director of the EAP specialization at the University of Maryland, in her article "Issues in International Employee Assistance Program Accreditation," emphasizes the dramatic growth and the development of the profession beyond the Employee Assistance Professionals Association (EAPA). She also discusses international EAP approaches to accreditation and identifies some of the potential cross-cultural limitations of American model EAPs. She describes the recent development of worldwide guidelines which have been sponsored by numerous EAP groups and suggests that these guidelines might be a pathway or intermediary step to accreditation for those international EAPs that may not be prepared to undergo formal COA accreditation.

This volume on EAP accreditation concludes with an article, "The Future of Credentialing and Accreditation in Employee Assistance Programs," written by current EASNA president Louise Hartley and EAPA president Don Jorgensen. These two EAP organization leaders examine future issues facing employee assistance programs and discuss the value and relevance of both program accreditation and individual practitioner certification.

R. Paul Maiden, PhD
Editor
University of Central Florida

Employee Assistance Program Accreditation: History and Outlook

Edward J. Haaz
John Maynard
Stephen C. Petrica
Charles E. Williams

SUMMARY. Accreditation is a means of verifying the professional competence and programmatic integrity of an employee assistance program (EAP). This paper examines the history of the accreditation of

Edward J. Haaz, MEd, CAC-Diplomate, is a principal of Mental Health Consultants, Inc., Furlong, PA. John Maynard, PhD, CEAP, is affiliated with SPIRE Health Consultants, Inc., Boulder, CO. Stephen C. Petrica, MDiv, MPH, is affiliated with the CDM Group, Inc., Chevy Chase, MD. Charles E. Williams, MHS, CEAP, is affiliated with the Center for Substance Abuse Prevention/Substance Abuse and Mental Health Services Administration, Rockville, MD.

Address correspondence to: Charles E. Williams, CSAP/SAMHSA, Rockwall II Building Room 920, 5515 Security Lane, Rockville, MD 20852 (E-mail: cwilliam@ samhsa.gov).

The authors gratefully acknowledge CSAP/SAMHSA for supporting the development of this paper.

The opinions expressed herein are the views of the authors and do not reflect the official position of their companies or of CSAP, SAMHSA, or the United States Department of Health and Human Services.

[Haworth co-indexing entry note]: "Employee Assistance Program Accreditation: History and Outlook." Haaz, Edward J. et al. Co-published simultaneously in *Employee Assistance Quarterly* (The Haworth Press, Inc.) Vol. 19, No. 1, 2003, pp. 1-26; and: *Accreditation of Employee Assistance Programs* (ed: R. Paul Maiden) The Haworth Press, Inc., 2003, pp. 1-26. Single or multiple copies of this article are available for a fee from The Haworth Document Delivery Service [1-800-HAWORTH, 9:00 a.m. - 5:00 p.m. (EST). E-mail address: docdelivery@haworthpress.com].

http://www.haworthpress.com/web/EAQ
Digital Object Identifier: 10.1300/J022v19n01_01

EAPs in the United States and Canada by the two dominant professional associations in the field, and makes some observations about the outlook for EAP accreditation. The two professional associations, driven by divergent philosophies, have evolved differently in their approach to accreditation. However, they share the conviction that control of standards is essential to the self-definition of a professional field, and has implications as well for marketing and governmental regulation. Accreditation thus has an important role in those areas, and should define acceptable standards in the emerging employee assistance environment, which entails such issues as managed behavioral health care, work-life, and international programs. Accreditation may also help advance thinking about current tensions in the field, and thus help shape its future. *[Article copies available for a fee from The Haworth Document Delivery Service: 1-800-HAWORTH. E-mail address: <docdelivery@haworthpress.com> Website: <http://www.HaworthPress.com> © 2003 by The Haworth Press, Inc. All rights reserved.]*

KEYWORDS. Accreditation, CARF, CEAP, COA, EAPA, EASNA, employee assistance, managed behavioral health care, work-life

INTRODUCTION

A profession, classically understood, is "a calling requiring specialized knowledge and often long and intensive preparation . . . maintaining by force of organization or concerted opinion high standards of achievement and conduct, and committing its members . . . to a kind of work which has for its prime purpose the rendering of public service" (Lawyers Title Ins. Corp. v. Hoffman, 1994; Georgetowne Ltd. Part. v. Geotechnical Servs., 1988). Entry to professional practice is generally restricted, either by state licensure or by the certification of a competent body of peers in the field, or both. Although it is unusual among other professions, the institutions in which health and human services professionals practice (e.g., hospitals, outpatient programs, rehabilitation facilities, HMOs, and social service agencies) are themselves often accredited. As employee assistance has emerged as a field with its own body of theory, knowledge, and skills, criteria for its competent practice have also developed, both for individual professionals and for the organizations in which they work. In this paper, the authors examine the history of the accreditation of employee assistance programs (EAPs) in the United States and Canada by the dominant professional associations in the field, the Employee Assistance Professionals Association (EAPA) and the Employee Assistance Society of North America (EASNA). This review of the history will permit us to make some observations about the outlook for EAP accreditation.

Employee assistance emerged in the 1940s out of the occupational health field. The first services, known as Industrial Alcoholism Programs (U.S. Department of Health, Education and Welfare, 1971), had the humanitarian and pragmatic business goals of identifying poorly performing employees with alcohol problems, helping them find appropriate treatment, returning them to productive employment, and thereby strengthening company productivity. As practitioners observed that employee productivity could be impaired by a range of personal problems beyond alcoholism, the field broadened from constructive confrontation-based Occupational [Alcohol] Programs (U.S. Department of Health, Education and Welfare, 1974) to more comprehensive employee assistance programs (Wrich, 1974). This transition marked a significant change. Once led by successfully recovering alcoholics from all walks of life, the shift resulted in an increased number of "degreed professionals" being attracted to the field. These individuals (most frequently members of one of the mental health professions) had the training to assess and intervene in a variety of emotional and behavioral problems. As they developed their workplace practice, many of them broadened their focus from alcohol-specific problems, training, and policies, to address a wide range of employee problems. Thus, the move began toward the so-called "broad brush" employee assistance identity, to the genesis of separate employee assistance professional associations, and eventually to the development of EAP standards.

The development of two distinct professional associations therefore bespeaks the diversity among practitioners as employee assistance evolved into a recognized profession. This diversity is evidenced by the shift from EAPs being staffed primarily by people of various occupational backgrounds who entered the field in part because of their personal experience of recovery and their concern for alcoholic coworkers, to increasing levels of staffing by professionals with advanced mental health training but minimal training or experience with the "recovering community." The earlier occupational programs tended to concentrate on "troubled employees" with performance problems caused by alcohol or drug abuse. The broad-brushed approach to employee assistance services which evolved later (and which continues to evolve) encompasses an ever-increasing range of workplace performance issues. These important differences are rooted deep in the history of the field (CONSAD, 1999), and they have influenced the evolution of EAP program standards, organizational accreditation, and professional credentialing.

EAP ACCREDITATION IN EAPA

For over twenty years, EAPA (and its institutional forerunner, ALMACA) has been a leader in the establishment of meaningful standards of practice in the employee assistance profession. The diversity of settings within which EAP services are provided, the myriad of backgrounds brought to the profession by employee assistance practitioners, and the spectrum of skills required to deliver expected results have made this effort a challenge.

The organization was founded April 27, 1971, as the Association of Labor and Management Administrators and Consultants on Alcoholism (ALMACA, the name being changed to EAPA in 1989), and received initial support from the National Institute on Alcohol Abuse and Alcoholism (NIAAA). ALMACA began the groundwork for accreditation in 1978 and 1979, a process that reached a turning point in 1981. That year the first Standards for Employee Alcoholism (or Assistance) Programs were drafted by a committee representing ALMACA, the National Council on Alcoholism, NIAAA, the Occupational Program

Consultants Association, and the American Federation of Labor-Congress of Industrial Organizations (AFL-CIO).

The employee assistance field grew significantly during the 1980s. In 1985, Drs. Paul Roman and Terry Blum published a paper (Roman & Blum, 1985; Roman, 1991) in which they identified six components of the EAP "core technology" (see Appendix A). According to Roman and Blum, these six functions constitute the necessary central activities of an EAP and combined they define the unique difference between EAPs and other workplace, self-help, or professional initiatives. Recognition for individual practitioners was planned in 1985 and formalized in 1987 with the establishment of the Employee Assistance Certification Commission (EACC). The EACC is an autonomous body created by EAPA to administer the Certified Employee Assistance Professional (CEAP) credential. The first CEAP examination was held in 1987, and the credential was awarded for the first time that year. In 1988, the ALMACA Board of Directors adopted a specific definition of an EAP:

> An EAP is a work-site based program designed to assist in the identification and resolution of productivity problems associated with employees impaired by health, marital, family, financial, alcohol, drug, legal, emotional, stress or other personal concerns which may adversely affect employee job performance.

Also in 1988, the ALMACA Board formed the Program Standards Committee to update and revise the 1981 standards, incorporating the new definition and the core technology. The name of ALMACA was formally changed to EAPA in 1989.

The EAPA Standards Committee began to issue a series of documents. "EAPA Standards for Employee Assistance Programs" (1990) set forth program standards, organized into six functional areas: program design, evaluation, implementation, management and administration, direct services, and linkages. In 1992, the "EAPA Standards for Employee Assistance Programs, Part II: Professional Guidelines" was published. This document added essential and recommended components to the 1990 standards.

From its earliest days, the EAPA Standards Committee recognized that market forces often resulted in services being sold as EAPs that did not meet the accepted standards of the profession. Discussion focused on the possibility of developing a program accreditation process based on the EAPA Standards, and in 1992 the Program Accreditation Subcommittee was formed within the Standards Committee to explore this issue. The subcommittee (which later became an EAPA standing committee) began work on the "EAPA Self-Administered Assessment Form for EAPs," which was published in 1994. The "EAPA Glossary of Terms" was also published that year. The publication of "EAPA Guidelines for International EAPs" in 1996 was the culmination of efforts by representatives from 14 countries to develop employee assistance guidelines applicable in countries and cultures worldwide.

Also in 1996, the EAPA Standards Committee began a comprehensive review of the "EAPA Standards and Professional Guidelines." The resulting revision, published in 1999, reflected important developments in the field. It provided guidance on issues that were potential sources of confusion, or that were important for differentiating acceptable from unacceptable EAP practices. The 1999 edition, incorporating an updated definition of an EAP, was organized into seven major sections: program design, management and administration, confidentiality and regulatory impact on protective rights, EAP direct services, Drug Free Workplace/Substance Abuse Professional direct services, strategic partnerships, and evaluation.

Meanwhile, the State of Florida Occupational Program Committee (FOPC) was developing an accreditation process for EAPs in Florida, assisted by Donald F. Godwin, former Chief of the Occupational Program Branch of NIAAA. The Workplace Research Branch of the National Institute on Drug Abuse awarded a contract to FOPC in 1989 to support a field test of the new process. In 1990, the EAP at the Honeywell plant in Tampa became the first program to be accredited by the FOPC. In 1991, EAPA reviewed the FOPC protocols for their usefulness as a national model, and decided that their best understanding would come from firsthand observation of the protocols being applied. Members of the EAPA Accreditation Committee went to Florida in

1993 to be trained as FOPC site reviewers and to participate in accreditation site visits at two programs. By doing so they were also pilot testing the FOPC protocols for possible adoption by EAPA. Their training and on-site experience were supported by the Workplace and Prevention Branch of CSAP, of which by that time Don Godwin had become Chief.

To build on the insights gained from this experience, the Accreditation Committee held informational discussions with five national accrediting bodies: the Joint Commission on Accreditation of Healthcare Organizations (JCAHO), the National Committee for Quality Assurance (NCQA), CARF (Commission on Accreditation on Rehabilitation Facilities), the Rehabilitation Accreditation Commission, American Accreditation Programs, Inc., and the Council on Accreditation for Children and Family Services (COA). The Accreditation Committee held an intensive two-day meeting in early 1994 to discuss what it had learned and to explore options for moving forward. Their deliberators considered the following.

Should EAPA Continue to Pursue Accreditation?

An accreditation process becomes important when the purchasers of professional services are limited in their ability to determine whether the services meet appropriate standards. Accreditation provides assurance that knowledgeable professionals have reviewed the services and found them to meet applicable standards. Accreditation is therefore the logical extension of a standards development process. Once standards are agreed upon, the accreditation review determines whether a particular entity has successfully operationalized them in its services.

EAPs thus fit the profile for accreditation. An EAP is a set of services for which professional standards have been developed, but most corporate and individual customers of EAPs don't have the means to determine whether any individual program meets those standards. At times, the term "EAP" may be used inappropriately to refer to sets of services that clearly do not meet EAPA standards. Since the term is unprotected, there is currently no way to prevent this from happening. Accreditation, then, would help define employee assistance practice and distinguish true EAPs from other sets of services that do not meet the standards of the profession. The process of accreditation provides a template to improve EAP services, and may help demonstrate to organizational decision makers why a particular activity is important and why staff time, expertise, and resources need to be allocated to it.

On the other hand, at the time there was reason to believe that in the absence of a viable accreditation process, government might take responsibility for defining employee assistance programs and standards. By 1994, when the Standards Committee met, this was already underway in several states and in federal regulations. Lawmakers do not have the time to become as knowledgeable about employee assistance issues as might be hoped, and they necessarily respond to political pressures. To avoid ill-conceived regulation and to maintain control of the profession the Committee concluded that it was in the best interest of the EAP field to develop a meaningful accreditation process.

What Actually Should Be Accredited?

EAPA Standards define an EAP as a work site-based program. The individual standards include some items that are properly the responsibility of EAP professional staff, while other items are the proper responsibility of the work site, that is, the host organization. Still others can only be achieved jointly by employee assistance professionals and elements of the host organization. An external EAP vendor may well provide services to one organization in which the EAP is fully integrated and operating according to the standards, while at the same time providing services to another organization in which the EAP fails to meet standards. The quality of any EAP, and its compliance with EAPA standards, is therefore a product of the interaction between the professional staff and the host organization. Ideally, accreditation should apply, and be awarded, to the specific programs arising from the joint responsibility and interaction of the vendor and host organization. As a practical matter, however, this may not be feasible for vendors operating EAPs in multiple host organizations. Therefore, the Committee considered that the external vendor organization itself, or the overall internal program, would likely be the entity to be accredited. To achieve such accreditation, vendors should be able to demonstrate their full integration into the workplace at a randomly selected sample of their multiple organizational clients. Multisite internal programs should be able to demonstrate full integration at a randomly selected sample of their sites.

What Will Accreditation Cost?

EAPs in North America are found in a huge array of settings. There are union-based programs, consortia of small groups and employers, internal programs in single and multisite companies, external programs serving companies of all sizes and in all geographic locations, and hybrid programs. Developing an accreditation process that is thorough enough to be meaningful yet inexpensive enough to be feasible in all employee assistance settings is, thus, always a challenge. The Committee directed that accreditation must not be allowed to favor larger vendors over smaller ones, nor must it inhibit innovative solutions to workplace problems, as long as the relevant standards are met.

Is There a Market for Accreditation?

As soon as it became available, EAPA sold over 300 copies of its "Self-Administered Assessment Form for EAPs." By March 1994, 35 organizations from around the world had made initial contact with EAPA about becoming accredited. Clearly, then, interest existed in the market even before a process was established. The Committee expected that demand for accreditation would increase sharply once the first programs were accredited, simply from the need of other programs to remain competitive in the EAP marketplace.

Should EAPA Be the Accrediting Body?

EAPA is strongly committed to retaining control of the standards for EAPs. The development of the EAPA Standards has been and continues to be an exhaustive and inclusive process that has resulted in significant progress in defining and distinguishing the employee assistance field. If EAP accreditation were to be based on standards other than EAPA's, members of the Association could ultimately lose control of their own profession and of the direction of the field. Another body with little EAP expertise could be defining the employee assistance profession. This could be especially harmful to employee assistance practice, given the potential for other accrediting bodies to lose sight of the fundamental importance of EAPA's workplace focus. Therefore, EAPA's ownership and control of the standards remained a "bottom line" as the Committee examined the options for moving forward with accreditation. The Committee carefully considered three options: developing the accreditation process internally within EAPA; supporting an external process outside of EAPA; or creating a joint venture with an existing recognized accrediting body.

Given its fundamental requirement that the EAPA Standards should remain the basis for accreditation, the Committee identified a number of significant advantages to collaborating with an outside accrediting body. Existing accrediting groups have the infrastructure and staff resources to develop and manage an accreditation process, while EAPA did not. Nor did EAPA have the resources to develop such an infrastructure. Using an outside group was also seen as protecting EAPA from potential liability associated with accreditation. External accrediting agencies could bring more credibility and objectivity to the process because they are independent, their process being less likely to be unduly influenced by internal EAPA considerations. At the same time, they bring already existing connections for lobbying federal and state lawmakers, and they already have working relationships with regulatory agencies. The Committee's challenge would be to find a suitable accrediting body with the flexibility and willingness to respond to EAPA's need to maintain ownership of the EAP Standards, and to help with such difficult issues as making accreditation financially feasible for all eligible EAPs.

Several forces were active as the Committee considered its options. Market demand, resistance to regulation, and ownership of standards drove the process inward toward EAPA. The expertise of existing accrediting agencies drove the process outward toward an external body. After much deliberation, the Committee decided unanimously to recommend the joint venture option. Under this plan, EAPA would select a qualified accrediting body with which to partner. EAPA and its partner would together create a joint commission for EAP accreditation with a majority of its members chosen by EAPA. The joint commission would approve the selection, training, and procedural guidelines for site reviewers/surveyors. EAPA would retain ownership of the content of the standards used in the accreditation process. On this basis, in July 1994 the Committee sent a "Request for Information" to the five accrediting bodies to solicit their feedback and interest in EAPA's proposed model. Two agencies, the Council on Accreditation (COA) and the Commission on Accreditation of Rehabilitation Facilities (CARF), responded to the Request with detailed proposals.

After further discussions and presentations from both of these groups, the Committee unanimously selected CARF as the accrediting body offering the best match for EAPA. This decision was based on a number of points, including CARF's openness to considering the joint venture design, its willingness to work with EAPA program standards, the relatively lower cost of CARF accreditation, the recognition and acceptance of CARF accreditation by the Joint Commission on Accreditation of Healthcare Organization (JCAHO) and National Commission on Quality Assurance (NCQA), and the fact that CARF's other standards tended to be more process oriented than prescriptive–a style consistent with EAPA's own approach to standards development.

The year 1995 was one of communication and negotiation. EAPA's chief operating officer and its president, who at that time were Sylvia Straub and George Cobbs, respectively, had both been personally involved in the selection process and subsequent discussions with CARF, and had been keeping EAPA's Board of Directors informed of the Committee's work. Debra Reynolds, chair of the Accreditation Committee, described the Committee's process and formally presented its recommendations to the Board at its spring 1995 meeting. George Cobbs sent letters to all EAPA chapter presidents to inform them of developments and to solicit chapter feedback. The Committee published an update on its work in the *EAPA Exchange* magazine, distributed a "fax-back" memorandum to the EAPA membership to stimulate awareness and discussion, and held a forum at the 1995 Annual Conference to gather additional input and encourage more dialogue. Meanwhile, EAPA and CARF continued to refine their proposed relationship and its related costs.

Despite these efforts, rumors began to flourish feeding concern of the accreditation process. Some internal programs–both labor and management based–worried that they would not be able to meet the accreditation criteria, and would be replaced by external programs. Some smaller external vendors were concerned that larger vendors would gain an advantage if accreditation existed. Ironically, the original intent was not to favor a specific EAP, and indeed to provide added support for highly integrated internal programs and for smaller vendors. Nevertheless, these concerns among others began to raise doubts among certain board members as to whether EAPA should continue to move forward with accreditation.

John Maynard and Debra Reynolds, representing the Accreditation Committee, met with the Board at its spring 1996 meeting to address the concerns. Following the Committee's presentation, the Board appointed a special Task Force to work with the Accreditation Committee "to develop a draft agreement with CARF by July 31, 1996." The Task Force, however, stated from the beginning that it was not in favor of pursuing accreditation, and, ultimately, it submitted motions to the Board that the negotiations between CARF and EAPA be terminated; that EAPA pursue state licensure initiatives instead of accreditation; and that the EACC (the CEAP credentialing subsidiary of EAPA) take over development of a low-cost quality assurance mechanism for EAPs. Although the Accreditation Committee requested that the Board poll the EAPA membership on whether to pursue accreditation, they declined to do so, and in July 1996 the Board approved the motion of the Task Force. The EACC declined to take responsibility for what it saw as an unworkable quality assurance mechanism. All accreditation efforts were then abandoned by EAPA, and the Accreditation Committee was dissolved. This remarkable turn of events was brought about in part because of uncertainty on the part of many about the impact of accreditation on individual employee assistance practitioners and their organizations.

Despite the abrupt end of EAPA's involvement, the forces that made accreditation desirable remain cogent. CARF has proceeded on its own to develop an EAP accreditation, using standards fully consistent with those of EAPA, while EASNA has joined with COA to develop its own EAP accreditation process. Yet for the employee assistance field's viability as a profession, it would seem essential that EAP accreditation be based on a single set of generally accepted standards. Many of its members continue to maintain that EAPA's standards, which have been refined over time by the experience of hundreds of employee assistance professionals providing services in every conceivable workplace setting, should remain the basis for EAP accreditation, regardless of the accrediting body.

CARF AND EAP ACCREDITATION

Despite the withdrawal of EAPA before their proposed partnership was finalized, CARF continued to develop a program of EAP accreditation, which became available in 1998. Within two years, some 20 programs had been accredited. CARF assumed EAPA's standards as the starting point for its own standards, and CARF's program reflects those origins. Consistent with the EAPA ethos, for example, CARF builds on the "core technology" tradition, and recognizes the CEAP as the fundamental credential for employee assistance practitioners (N.K. Magas [Managing Director, CARF Behavioral Health Customer Service Unit], personal communication August 29, 2003). However, diverging from at least one projection of EAPA's earlier Accreditation Committee, CARF does accredit individual components of larger EAP programs (e.g., one location of a multisite internal program, or a single program run by external employee assistance contractor).

Although they remain consistent with EAPA's standards, CARF's accreditation criteria have evolved independently. CARF periodically convenes a National Advisory Committee of purchasers, providers, and consumers of EAP services to review the standards. Potential changes are then submitted for field review before their adoption is approved by CARF's Board of Trustees (Magas, 2000).

EASNA STANDARDS AND ACCREDITATION

Founded by members of both EAPA and the Canadian Employee Assistance Program Association, EASNA has focused since it inception in 1983 on the need to develop the highest standards of professional practice for employee assistance as opposed to occupational alcohol programs. It has had a binational character from the beginning, with members in both the United States and Canada, and the two nationalities represented equally on its Board.

In the mid-1970s, Wayne Corneil witnessed the evolution of the alcohol treatment field in the United States from dual vantage points, both in Canada and the U.S. As a doctoral student at the Johns Hopkins University School of Hygiene and Public Health (supported in part by an NIAAA training grant), and as a staff member of Health Canada (the Canadian federal health ministry), he was involved in regulating health care in various business sectors, including the Canadian airlines. An initiative by the Canadian Labour Congress (Canada's counterpart to the AFL-CIO) started a dialogue regarding the need for national standards for helping troubled employees. The specific concern in this case was for airline pilots who were being evaluated for alcoholism, receiving treatment, and subsequently returning to the workplace. Valid and reliable assessments were needed. A two-week training program was designed through an initiative by Health Canada, and offered to peer professionals through the community college system. This training was an initial effort to provide workplace professionals with standardized training in the assessment of alcohol-related problems in troubled employees.

In 1977, Dr. Corneil attended an ALMACA meeting in Detroit, where he spoke about the need for an international component within the organization. After discussing trends north of the border, he suggested that the ALMACA Board consider the need for program standards and a certification process for their members. He argued that these peers made decisions that affected the fate of many alcoholics and their coworkers. In 1978 and 1979 ALMACA began working to devise program standards for the EAP field, an initiative that came to fruition in 1981, as described earlier. Observers of the EAP field in both Canada and the United States followed these developments closely.

In 1982, Keith McClellan, Wayne Corneil and George Watkins met informally, and discussed the need for degreed professionals in the field of EAP. The discussion centered on traditional occupational alcoholism programs, core technology issues, the history of ALMACA and the evolving "broad brushed" approach to EA. The three decided to continue their discussion in September 1982 at the Input Conference in Toronto, Ontario. The Toronto attendees agreed on the desirability of a new association specifically for "Employee Assistance" professionals. It was at that conference that the "EASNA" name was conceived, with the mission of the organization defined to produce a body of work aimed at minimal program standards and an accompanying accreditation process. These standards were to be premised on the broad brush approach to EA, with the goal of defining more clearly the professional roles and activities of practitioners working in the field. One of the Toronto attendees, James T. Wrich (who was one of the original 100 NIAAA-funded State Occupational Programs Consultants [the so-called "Thundering Hundred"] and who also served on the ALMACA Standards Committee), was later recruited by the group to explore how far these priorities could be advanced within the ALMACA framework. (As described earlier, ALMACA did move to institute certification for individual practitioners but was markedly more reticent to accredit programs.)

In January 1983, a small core group of participants in the 1982 Input Conference took the initiative to draft bylaws and establish EASNA as a nonprofit organization, with the goal of providing "a leadership role in the encouragement of quality EAP services through the development of Program Standards and an Accreditation process." As leaders of the newly founded EASNA, they struck an agreement whereby the Performance Resource Press, Inc. (PRP, of which George Watkins was president) would sponsor the fledgling organization for three years, during which EASNA would begin to develop its organizational identity and membership in both the United States and Canada. After three years, it was agreed, EASNA would survive on its own. EASNA was also eventually incorporated in Canada, and so became a binational organization.

At the 1983 North American Congress on Employee Assistance Programs in Dearborn, Michigan, attendees again discussed the lack of standards by which to define the field of EA. Concerned that some programs could provide ineffective–perhaps even harmful–services to employees and their companies, there was marked interest in the question of standards.

In 1984, its first year of existence, EASNA promulgated the EAP field's first code of professional ethics. That same year the new organization also partnered with The Haworth Press, Inc., to begin publication of its refereed journal, the *Employee Assistance Quarterly*. The necessary groundwork was thus being laid for the creation of standards, and EASNA spent the next five years developing the body of work necessary to accomplish its goals.

Dating back to 1983 the field had for the first time an organization focused solely on promoting professional EAP practice, standards, and, ultimately, accreditation. After three years, as planned, EASNA separated from PRP and the North American Congress to become a fully independent entity, and the organization was poised to develop standards. By the spring of 1989, EASNA had generated sufficient interest to convene a conference on the creation of program standards, and so initiated its own annual professional development institute. EASNA's First Annual Institute was held June 18 to 22, 1989, at Allgauer's Hotel in Chicago. President Wayne Corneil stated in his opening address to participants:

> Governments are moving to regulate the employee assistance field. Who will set the standards, the regulations to be enshrined in

legislation? There are those who would have regulations spawn a whole new industry of private agencies controlling quality assurance. Who will regulate these groups? Will a few well-connected individuals determine how we are to be measured or will the collective wisdom of practitioners form these regulations?

EASNA believes that this issue will determine not only the future but also the very existence of EAPs. It is essential that everyone from our field, not just a select few, have the opportunity to contribute to the formulation of standards. I trust that you will take advantage of the session here to participate fully in this important task.

EASNA is a leader in the employee assistance movement. The reason it has been in the forefront is the active involvement of its members. I am confident that your contributions over the next few days will again advance our chosen field. I am certain also that you will find the interaction with your colleagues stimulating and challenging, adding to your professional expertise. (Haaz, 2002)

As Corneil noted in his keynote address, government interest in EAPs had increased, and governmental regulation was possible. The emphasis of the conference was therefore to facilitate a *peer-driven* process of creating standards that would lead to an eventual accreditation process. The next few days found the 89 participants (Appendix B) sequestered in smaller work groups for 12 to 14 hours a day under the guidance of EASNA's Board of Directors. By the end of that first Institute, the work groups, taking notes on flip charts, had generated over a hundred sheets of newsprint, which several editors were recruited to collate and distill. Over the course of several months, these editors submitted their material to McClellan and Corneil who, with other key participants, produced a first draft of the EASNA Standards.

In the fall of 1989, the first draft was sent out to the field for comment, and responses were received from over a hundred employee assistance professionals. After several revisions, this peer-reviewed, consensus-driven process resulted in the formulation of the "EASNA Standards for EAPs." Published in 1990, these first standards require that an EAP must be easily accessible; "user-friendly" for consumers; respectful of confidentiality; centered on the employee; focused on prevention; and staffed by competent professionals.

While the Standards Committee worked on refining the standards, the Accreditation Committee was hard at work developing a process by which programs could be measured against these standards. Many individuals, led by very capable chairpersons (Appendix C), worked countless hours to refine the standards and the process, and by early 1991 the accreditation process was instituted. The first program in the U.S. to be EASNA-accredited was Genesis EAP in Iowa, and CanCare in Ontario became the first Canadian EAP accredited soon thereafter. This established a momentum that was quickly embraced by the Canadian employee assistance community but somewhat more slowly by EAPs in the U.S.

There were many developments between 1991 and 2000. During that time there were three revisions to the Standards. All revisions took place in an open forum and with input by many people (including members of both EASNA and EAPA), engaged in all aspects of the EAP field. Many programs in the U.S. (both internal and external) as well as most of the major Canadian EAP organizations began to fully support the standards and accreditation process. EASNA accreditation achieved a high profile by coming to be specified in Requests for Proposals from increasing numbers of Canadian purchasers. This was a pivotal point in the evolution of the process. Now, in order to compete in the Canadian marketplace, EASNA accreditation became essential. Within these ten years, most large and mid-sized Canadian EAPs, and several mid-sized American programs, were involved in the accreditation process. Of note is that the internal EAP of a large binational company, the Bank of Montreal, also became accredited.

EASNA, it might be said, became a victim of its own success. The task of refining the standards and meeting the demand for site visits started to become daunting, even for the dedicated volunteers committed to the work. Therefore, in 1999 the EASNA Board authorized an exhaustive search for a partner to carry accreditation to the next level of success. EASNA sought collaboration with EAPA as well as with a number of accrediting organizations, and ultimately found its partner in the nonprofit Council on Accreditation (COA). Founded in 1977, COA is an international organization that originally accredited child welfare programs. Their experience developing EAP standards dates back to 1987, when they began to accredit multiservice organizations that provided EAP services as one component among others. In 1999, EASNA entered an agreement with them whereby COA would administer the EASNA accreditation process for stand-alone EAPs, including internal and external programs, and those operated by managed behavioral health care organizations and work-life organizations.

Soon after, the Center for Substance Abuse Prevention (CSAP) and the Center for Mental Health Services, both of which are part of the U.S. federal government's Substance Abuse and Mental Health Services Administration, supported a preliminary review of the work on standards done by EASNA, COA, EAPA, the federal EAP guidelines, and others in the field, with the goal of identifying best practices in EAPs. Masi Research, Inc. (the workplace consulting firm led by Dale Masi, DSW) became the house consultants for COA, and teamed with the EASNA and COA Standards Committees to review, compile, sort, and integrate all appropriate standards, guidelines, and comments. They were also to identify and develop new standards as necessary. They ultimately created a single combined set COA and EASNA standards, with a glossary of terms and other tools for conducting the accreditation process. By the summer of 2000, four EAPs in the United States and one in Canada served as "beta sites" for the revised accreditation procedures and standards. This entailed their undergoing the full accreditation process from self-study to site visit to final accreditation decision. Subsequently, the wider EAP field was approached for continued comment and input. The result of this process was the *EAP Standards and Self-Study Manual, 1st Edition*, which was officially released by COA in 2001. This volume reflects input from the organizations, comments from the field, the COA accreditation format, and the recommendations of the beta site reviewers.

EASNA and COA renegotiated their contract in 2002. Under the terms of the new agreement, EASNA is now officially a supporting organization of COA, while COA has become a stand-alone EAP accreditation agency. COA "owns" the standards and process, and selects and trains peer reviewers, while it is "advised" by the EASNA Standards Committee. Senior EAP professionals serve on the COA accreditation commission. The process is completed in four phases: Application; Self-Study (the first two completed by the organization seeking accreditation); Site Visit (conducted and reported by a team of Peer Reviewers); and finally, the Accreditation Decision made by the COA Accreditation Commission.

After another round of field comment and review by a panel that included members of both EAPA and EASNA, the second edition of the *Manual* was published in 2003 (Council on Accreditation). The new standards represent a consensus of the field, reflecting input from service providers, regulators, policy makers, the professional associations, academic researchers, consumers, and funders. A standards and accreditation process driven by the entire employee assistance field is therefore now in place. By its synergistic partnering with the accreditation experts at COA, EASNA has propelled the accreditation standards and process to a new level. To help implement the accreditation process, EASNA has long offered a formal mentoring program as a benefit to its organizational members seeking accreditation; and in 2003 it updated the mentoring program to make it consistent with the second edition of the *EAP Standards Manual*. The result of all this is a more sophisticated set of policies and procedures, which are part of an independent, standardized process.

THE OUTLOOK FOR EAP ACCREDITATION

Two agencies (CARF and COA) currently accredit EAPs. While at first glance this situation may seem anomalous, it is not unusual in health care and human services. For example, JCAHO and NCQA both accredit health care organizations, and JCAHO, CARF, and COA all accredit drug and alcohol treatment facilities. Nor is it necessarily undesirable; indeed, it may be understood as a manifestation of the field's interest in accreditation and recognition of the importance of certifying professional standards. Note that COA reported greater interest and response from employee assistance professionals than they expected when they undertook revisions to the first (2001) edition of their EAP standards manual.

The employee assistance professional associations are in broad agreement on the necessity of self-regulation, without which the profession is at risk of being regulated by bodies or forces that do not share its self-understanding (Claeys, 2000). Self-regulation has also been seen as increasingly important for maintaining the integrity of the field's identity with the growth of managed behavioral health care, work-life programs, and workplace issues raised by economic globalization. The question of what exactly is implied by "employee assistance" was identified as a problem by the late 1980s, and that problem was only exacerbated through the 1990s as increasing numbers of vendors began to offer different "employee assistance" and related products. In July 2003, EAPA's Board adopted a revised definition of EA: "Employee Assistance is the work organization's resource that utilizes specific core technologies to enhance employee and workplace effectiveness through prevention, identification, and resolution of personal and productivity issues."

Market forces thus challenge the employee assistance field to clarify exactly what constitutes an EAP, and the power to decide what is accreditable as an EAP is a formidable response to that challenge. At the same time, EAP vendors of all varieties in the competitive U.S. marketplace are beginning to see accreditation as a marketing asset, as Canadian EAP vendors already have. The competitive advantage may also come to be seen by internal programs and by smaller- and medium-sized EAPs. By being accredited their claim gains credence that smaller or internal EAPs can be every bit as good as large national providers.

With the advent of managed care as both a dominant force in health care in the United States and as a major vendor of employee assistance products, a new challenge is to differentiate from managed care in the understanding of the workers who are the end users of employee assistance services. Ethical issues are raised when traditional employee assistance functions (e.g., serving as either a gatekeeper or a referral source for substance abuse treatment or mental health services) conflict with organizational needs. The relationship between managed behavioral health care and employee assistance is too profound and complex to explore in this article, but the questions outlined here suggest the importance of careful attention to professional standards and accreditation criteria for the employee assistance field.

By maintaining standards that are defined by practitioners in the field (primarily through their professional associations), accreditation has the potential as well to uphold the definition of employee assistance as a highly integrated workplace-based technology to enhance employee and workplace effectiveness. The employee assistance field, operating on that understanding, has traditionally opposed including EAP services among human resource or mental health benefits. Accreditation standards, by reinforcing this key distinction, could thus be a useful tool for corporate benefits managers and consultants.

Reverberations of historical differences in the field persist, but they are presently being addressed. In November 2002, and January and April 2003, representatives of the EAPA and EASNA Boards convened for intensive discussions to move toward collaboration on a wide range of professional issues. As a result of these meetings, the joint EAPA/EASNA task force achieved consensus on the importance of individual certification and program accreditation. The parallel moves toward accreditation that the two organizations began in the 1980s, and that included earlier unsuccessful attempts to collaborate, are beginning to show signs of coalescing and converging.

Both EAPA and EASNA recognize the critical need to define and ensure the quality of employee assistance services. Thus, questions that have been raised as it has dealt with standards and accreditation point to larger tensions about its professional identity and self-understanding. Is it a field driven by professional principles or by such market forces as purchaser demand, vendor promotions, benefits management, and the recommendations of business consultants? Should employee assistance services best be understood as a professional practice or as a trade; as a human resource benefit or as a workplace technology? To the extent that more profound engagement with these questions is reflected in the process of developing standards and applying them in accreditation, the field will be strengthened and its foundations reinforced.

REFERENCES

Claeys, S. (2000, Fall). An opportunity for self-regulation. EAP Digest [online]. Available: http://www.prponline.net/Work/EAP/Articles/an_opportunity_for_self_regulatoin. htm [sic].

CONSAD Research Corporation. (1999). Historical overview and evolution of employee assistance programs. EAP Handbook [online] Available: http://www.consad. com/eap/chapter02.htm

Council on Accreditation. (n.d. [2003]). *A History of COA Accreditation.* New York: Author.

Georgetowne Ltd. Part. v. Geotechnical Servs., 230 Neb. 22, 430 N.W.2d 32 (1988).

Haaz, E. (2002, May). EASNA Standards and Accreditation History. Paper presented at the EASNA 14th Annual Institute, Quebec City, Canada.

Lawyers Title Ins. Corp. v. Hoffman, 245 Neb. 507, 513 N.W.2d 521 (1994).

Magas, N.K. (2000, Winter). Accreditation option: CARF . . . the Rehabilitation Accreditation Commission. *EAP Digest*, 15-17.

Maynard, J. (2002, May). The history of EAP accreditation within the Employee Assistance Professionals Association: 'A long and winding road.' Paper presented at the EASNA 14th Annual Institute, Quebec City, Canada.

Roman, P.M. (1991, July). Core technology clarification. *Employee Assistance*, 8-9.

Roman, P.M., & Blum, T.C. (1985, March). The core technology of employee assistance programs. *The Almacan*, 8-11.

U.S. Department of Health, Education, and Welfare. (1971). First special report to the U.S. Congress on alcohol and health. Rockville, MD: Author.

U.S. Department of Health, Education, and Welfare. (1974). Second special report to the U.S. Congress on alcohol and health. Rockville, MD: Author.

Wrich, J.T. (1974). The employee assistance program. Center City, MN: Hazelden Foundation. [Revised edition published in 1980.]

APPENDIX A

EAP Core Technology

1. Consultation with, training of, and assistance to work organization leadership (managers, supervisors, and union stewards) seeking to manage the troubled employee, enhance the work environment, and improve employee job performance; and outreach/education of employees/dependents about availability of employee assistance services;
2. Confidential and timely problem identification/assessment services for employee clients with personal concerns that may affect job performance;
3. Use of constructive confrontation, motivation, and short-term intervention with employee clients to address problems that affect job performance;
4. Referral of employee clients for diagnosis, treatment, and assistance, plus case monitoring and follow-up services; organizations, and insurers;
5. Assistance to work organizations in managing provider contracts, and in forming and auditing relations with service providers, managed care organizations, insurers, and other third party payers;
6. Assistance to work organizations to support employee health benefits covering medical/ behavioral problems, including but not limited to alcoholism, drug abuse, and mental/ emotional disorders; and
7. Identification of the effects of employee assistance services on the work organization and individual job performance.

APPENDIX B

Participants in the First EASNA Institute, 1989

Dorothy Agner	Robert F. Alford	Stacy Balonick
Samuel Berkowitz	Maxine Berry	Gary Bloker
Thomas Budziack	Frank Burger	Gail S. Buss
Peter J. Clark	Wayne Corneil	Jim Costabilo
Walter H. Cox	Chelle Dainas	Mary L. DePillars
Diane E. Dockery	Joanne Dougherty	J. Chip Drotos
Heyward L. Drummond	Donald E. Dufek	Yvonne Eichorn
John E. Fertig	Michael Fortin	Jack Freckman
Michael B. Garfield	Sewel Gelberd	Patsy Gillespie
Bill Graham	Edward J. Haaz	Paula Harkness
Cathy Hart	Steven Haught	Ray Johnston
Edward A. Kaczmarek	Dorothea Kaplan	Linda Karlovec
Harry W. Kelm	Linda Kirk	Leo Lalonde
Michel Legault	Diane LePage-Racette	Scott Lindsley
Vicki M. Lodge	Rita Losee	Pat Marchand
Katherine Mathias-Maher	Keith McClellan	Donna Menarek
Diana Meyers	Marjorie J. Middel	Mary Mitchell
Leslie Modrack	Lionel Moore	David O'Brien
Carol A. Pape	Donald Pare	Sandra Parsenue
Gerard Perusse	Barbara Pipes	Mary Pittman
Linda M. Poverny	Charles Rebault	Charlotte Schmidt
Roger Schneider	Dennis Schram	Carol Seacord
Gerald Shulman	Andrew M. Siegel	John Sizemore
David E. Smith	Sally Spritz	Charles E. Stanley
Susan Stanley	Ray Steinkerchner	Linda Stoerr-Scaggs
Harry Sundell	Cheryl Thomas	Eleanor Thomas-Grumbach
Michael Tinken	Carl Tisone	Joann Trullo
Jeanne G. Trumble	Fred Upshaw	Monique Quirion van Gent
Bonnie Warton	Celia Webb	Sanford Weinberg
Alistair Westgate	Kenneth Wolf	Dennis Wilson
Georgina Witcovic	James T. Wrich	R. John Young
Arthur Zaragoza		

APPENDIX C

EASNA Committee Chairpersons

Chairpersons of the EASNA Accreditation Committee during this period were:

Bill Graham (Canada)
James Offield (U.S.)
Edward Haaz
Barbara Marsden
Suzanne Claeys
Rita Fridella

Chairpersons of the Standards Committee were:

Sanford Weinberg
Sharon Pocock
Charles Williams
Donna Scotten
Marilyn Hayman
Barbara Marsden

Developing Standards for Accreditation

Stephanie Pacinella

SUMMARY. Standards development is a continuous process that relies on an inclusive, consensus-building methodology to ensure that standards maintain relevance in an ever-changing field. This article provides an overview of the framework for the Council on Accreditation (COA) Employee Assistance Program (EAP) Standards, and details the steps in the standards development process that resulted in both the first and current editions of the *COA EAP Standards and Self-Study Manual*. Areas of emphasis for future standards development are described. *[Article copies available for a fee from The Haworth Document Delivery Service: 1-800-HAWORTH. E-mail address: <docdelivery@haworthpress.com> Website: <http://www.HaworthPress.com> © 2003 by The Haworth Press, Inc. All rights reserved.]*

KEYWORDS. Accreditation, employee assistance programs, standards development

Stephanie Pacinella, MA, is Assistant Director of Standards Development and Performance Measurement, Council on Accreditation, 120 Wall Street, 11th Floor, New York, NY 10005 (E-mail: spacinella@coanet.org).

[Haworth co-indexing entry note]: "Developing Standards for Accreditation" Pacinella, Stephanie. Co-published simultaneously in *Employee Assistance Quarterly* (The Haworth Press, Inc.) Vol. 19, No. 1, 2003, pp. 27-34; and: *Accreditation of Employee Assistance Programs* (ed: R. Paul Maiden) The Haworth Press, Inc., 2003, pp. 27-34. Single or multiple copies of this article are available for a fee from The Haworth Document Delivery Service [1-800-HAWORTH, 9:00 a.m. - 5:00 p.m. (EST). E-mail address: docdelivery@haworthpress.com].

Digital Object Identifier: 10.1300/J022v19n01_02

INTRODUCTION

The Council on Accreditation (COA) first released standards, designed for multiservice organizations providing employee assistance program (EAP) services, in 1987. In 1999, COA partnered with the Employee Assistance Society of North America (EASNA) and created a separate accreditation product specifically tailored for internal and external EAPs. The standards that embody that product, the *Employee Assistance Program Standards and Self-Study Manual, 1st Edition*, were written explicitly for stand-alone employee assistance programs. The impetus for this separate product arose from a need in the field for standard practice guidelines to help describe a well-functioning EAP. At the time of publication the field was largely unregulated, and many groups fashioned themselves as EAPs but did not exhibit the components that define a true EAP. The Employee Assistance Professional Association (EAPA) terms those essential components the EAP Core Technology.

The COA EAP Standards (see Appendix) are structured to encompass the EAP core technology and the process requires organizations to have those key elements in place to achieve accreditation. The standards that address the core technology elements provide a standardized framework for providing services that support positive outcomes for clients, and encompass the values that are important in the EAP industry. Key areas of the framework are employee education and outreach; information and referral, and assessment and referral services; training for supervisors and union representatives, management/supervisory consultation; work-life services; short-term counseling; organizational development; critical incident stress management; and drug-free workplace services. Additional standards address areas vital to well-functioning organizations and services, including governance and administration; management of EAP human resources; health and safety; legal liability; quality improvement; contracting for services; personnel and affiliate competence; confidentiality and privacy protections for clients; conflicts of interest; and ethical considerations.

An open, democratic process is used to develop a product that incorporates these values and represents best practices in the EAP industry.

The emphasis lies in a continuous consensus-building process to ensure that standards maintain their relevance in an ever-changing field. The COA standards development process includes several steps, as outlined below. Those steps, at times, may overlap, and on occasion may change order depending on the issue being addressed. The completion of a new book of COA standards generally spans two to four years from the initiation of revision to publication, depending on the scope and nature of revision and the extent of new standards development.

DATA COLLECTION

COA collects data from various sources on an ongoing basis to inform and support its standards development process. These sources include feedback from accredited organizations and peer reviewers that conduct accreditation site visits; observations made by staff during visits to organizations; review of current research and survey data; information gathered at national conferences; input obtained from national membership associations; and the analysis of accreditation data. This information gathering helps staff to identify the standards that need revision or further development, as well as areas of practice for which new standards need to be developed. Throughout the development of COA's EAP Standards, COA received guidance from several leading EAP industry groups including EASNA, EAPA, the EAP Roundtable, and the EAP Joint Industry Alliance.

LITERATURE REVIEW

At the beginning of the process of developing a new book of standards there is a period of time that allows for an intense focus on researching and locating resources, and reviewing the literature that has been gathered; thereafter, literature reviews are an important, ongoing staff responsibility. During this time it is also useful to look at descriptive program material and to visit representative programs throughout the United States and Canada to observe organizational practices as they occur in the field.

STANDARDS ADVISORY PANELS

COA engages experts in various fields of practice to review the standards and provide input throughout the standards development process. These experts are grouped together by field of practice and serve on one of several permanent, ongoing standards advisory panels. Panel members consist of representatives from the field, including agencies, peer reviewers, board members, sponsoring and supporting organizations, policy and research organizations including colleges and universities, accreditation commissioners, and other relevant groups such as governmental entities, funders, and managed care organizations. In support of this vital consensus-building process, COA established a panel specifically to represent the employee assistance field.

The EAP Standards Advisory Panel represents a full range of perspectives from across the EAP field, including internal, external, large, small, private, public, American and Canadian EAPs. Approximately fifty stakeholders representing those various EAP models attended the first EAP Standards Advisory Panel meeting in April 2002 to impart their knowledge and perspectives on EAP best practices.

The initial panel meeting occurs early in the process to allow panel members the opportunity to provide their feedback on the current standards and to solicit input on what needs to be revised or newly developed (Figure 1). Suggestions provided by the EAP Standards Advisory Panel for revision of the *EAP Standards and Self-Study Manual, 1st Edition*, focused on further development of standards for work-life services, online and telephone services, public model EAPs, and international EAPs, as well as the need for the separation of standards for affiliate providers from standards for staff members. Typically, panels will meet three times during the standards development process: prior to standards drafting, upon completion of the first draft of the standards, and again following field review.

FIGURE 1. Standards Development Process

Data Collection and Literature Review

Standards Advisory Panel–*Meeting 1*

Standards Drafting

Standards Advisory Panel–*Meeting 2*

Standards Revisions

Field Comment

Analyze and Incorporate Field Comment

Field Testing

Analyze and Incorporate Field-Testing Data

Standards Advisory Panel–*Meeting 3*

Final Standards Revisions

Production

Publication

Standards Updates

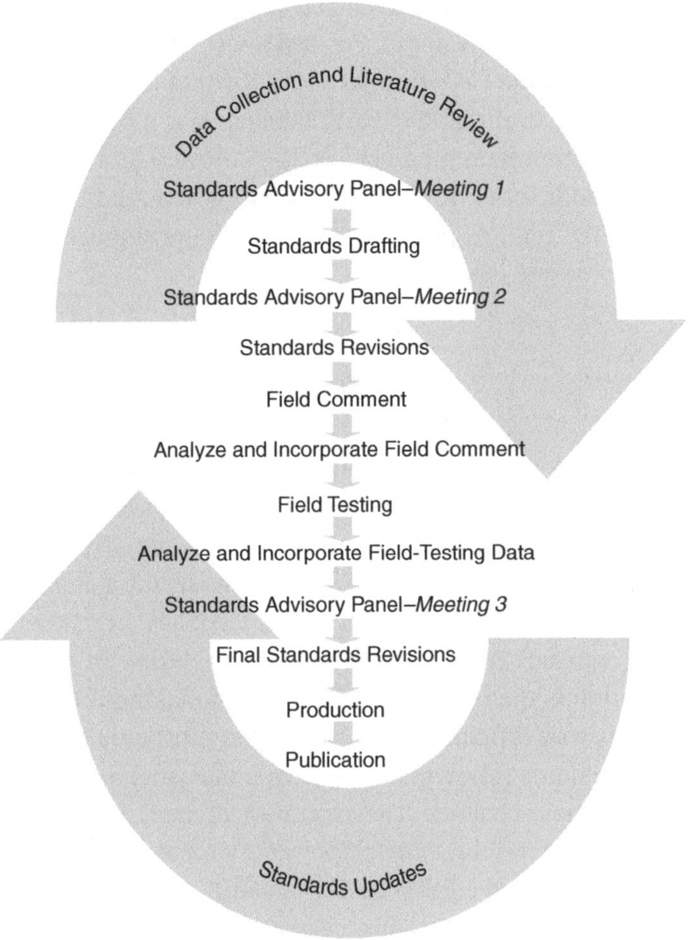

STANDARDS DRAFTING AND FIELD COMMENT

Although initial standards drafting occurs throughout the early phases of the standards development process, it is after the first panel meeting that staff work in earnest to develop the first draft of the manual. Following the panel's review of the first draft and incorporation of any changes, the draft is made available to the public for comment. Field review involves posting the manual for a specified period of time and notifying all agencies, peers, and other stakeholders of its availability and their opportunity to provide input. Postcards were sent to approximately 500 stakeholders in the EAP field to notify them of the EAP field comment period. The draft of the EAP manual was posted on COA's Website for one month (September 2002) to allow visitors the opportunity to download it for review and comment. When the field comment period ended, the field response was analyzed and standards were redrafted to incorporate input.

FIELD TESTING

Field testing is performed to promote reliability and validity in the accreditation process. It may be carried out with specific standards, new service sections, or the entire manual when feasible. This evaluative process involves the participation of agencies using the standards in the accreditation process. The goal is to obtain feedback from the peer reviewers and agencies regarding the utility and design of the manual. This information is then used to assist COA in making refinements to the standards before official publication. When substantial revisions are made to the standards following this process, the panel may be engaged again to review those changes. The 1st edition of the EAP Standards underwent a rigorous field-testing phase at five sites. The design and methodology of the 2nd edition EAP Standards remained constant, so another field testing phase was unnecessary; all revisions and additions were solicited through the field comment period.

PRODUCTION

Production is the final fine-tuning and editing stage of standards development. This is a detail-oriented process that spans several months. This is the last opportunity to review standards language, the consistency and flow of each section of the manual, and to make edits and adjustments to the overall product. The production stage involves formatting the document into a user-friendly layout, multiple proof readings, designing the artwork for the manual, and coordinating the separate documents that make up the entire self-study manual. The final product, in this discussion the *Employees Assistance Program Standards and Self-Study Manual, 2nd Edition*, published in Spring 2003, is a comprehensive blueprint of best practice principles. These standards are available through the Council on Accreditation.

STANDARDS UPDATES

COA's mechanism for updating standards allows revision of a standard in need of substantive change during the time when a accreditation manual is in current standing. For example, a standard update may occur because of a change in practice, the need for further clarification of a standard, or to provide additional flexibility to a standard. Updating a standard involves reviewing relevant literature on a particular issue, as well as consulting individuals both internal and external to COA. The information collected is reviewed and discussed thoroughly, and careful consideration is given in crafting the appropriate language for the standard update. Standards updates are produced on an as needed basis, and posted on COA's Website. In addition, copies of Updates are distributed to all organizations when they apply for accreditation, and to peers as part of the material they receive for a site visit. This process helps to respond to the field's needs and input on a more frequent basis outside of the lengthy manual revision process.

THE FUTURE OF STANDARDS DEVELOPMENT

COA maintains its course of standards development to further the evolution of each of the standards products. Focus for future standards development has pointed to evidence-based practices as a primary source of best practice principles. In addition, data collection increasingly will promote and support ongoing improvements in organizational capacity, as well as positive service delivery processes and outcomes in accredited organizations. As COA continues to strive to develop and strengthen accreditation standards relevant to the EAP field, it is imperative that stakeholders continue to participate in the process to help ensure a future of quality EAP services for clients who are entitled to the highest level of quality service.

REFERENCES

COA Standards and Self-Study Manual, 7th Edition, Version 1.1. Council on Accreditation: New York, NY, October 2001.

Employee Assistance Program Standards and Self-Study Manual. Council on Accreditation and the Employee Assistance Society of North America: New York, NY, June 2001.

Employee Assistance Program Standards and Self-Study Manual, 2nd Edition. Council on Accreditation: New York, NY, January 2003.

Employee Assistance Professionals Association. "EAP Core Technology" [Online]. Available: http://www.eapassn.org/public/pages/index.cfm?pageid=521

The Council on Accreditation Employee Assistance Program Accreditation Process

Timothy J. Stockert

SUMMARY. The Council on Accreditation Employee Assistance Program accreditation process is time limited and facilitative in nature. It is a distinct, step-by-step process that involves an internal and external review of an organization's policies, procedures, and practices based on standards of best practice. This article provides an overview of the steps in COA's EAP accreditation process from application to reaccreditation, and discusses several of the distinguishing features of the process. *[Article copies available for a fee from The Haworth Document Delivery Service: 1-800-HAWORTH. E-mail address: <docdelivery@haworthpress.com> Website: <http://www.HaworthPress.com> © 2003 by The Haworth Press, Inc. All rights reserved.]*

KEYWORDS. Accreditation, employee assistance programs, accreditation process

---

Timothy J. Stockert, MBA, MSW, is Manager of EAP Services, Council on Accreditation, 120 Wall Street, 11th Floor, New York, NY 10005 (E-mail: tstockert@coanet.org).

[Haworth co-indexing entry note]: "The Council on Accreditation Employee Assistance Program Accreditation Process." Stockert, Timothy J. Co-published simultaneously in *Employee Assistance Quarterly* (The Haworth Press, Inc.) Vol. 19, No. 1, 2003, pp. 35-44; and: *Accreditation of Employee Assistance Programs* (ed: R. Paul Maiden) The Haworth Press, Inc., 2003, pp. 35-44. Single or multiple copies of this article are available for a fee from The Haworth Document Delivery Service [1-800-HAWORTH, 9:00 a.m. - 5:00 p.m. (EST). E-mail address: docdelivery@haworthpress.com].

INTRODUCTION

Although there are many different accrediting bodies in the world today for numerous types of services/programs, there are two principal accrediting organizations for Employee Assistance Programs (EAPs). The Council on Accreditation (COA) and the Commission on Accreditation of Rehabilitation Facilities (CARF) are the two primary accreditors of EAPs in North America. Both organizations have similar processes for accrediting EAPs that require pre-site submission of documentation and on-site review by volunteer peers. Over the years, both organizations have refined, adapted, and altered their processes to meet the changing needs of the organizations they accredit. While COA and CARF have very similar processes, this article will focus on process information that is specific to EAPs seeking accreditation with COA.

COA'S FACILITATIVE PROCESS

COA describes itself as a "facilitative accreditor," meaning that it provides information and assistance to organizations throughout the accreditation process to help them interpret best practice standards and understand how these standards apply to the services they provide. COA is also facilitative in that it gives organizations multiple opportunities to show how they are complying with best practice standards (e.g., self-study, site visit, responses to reports, etc.), and additional time to implement processes that are not fully compliant with standards of best practice.

TIME FRAME

In general, the time frame for organizations to become accredited with COA is limited. New organizations seeking accreditation have two years to complete the process. Organizations seeking reaccreditation are notified eighteen months in advance of their expiration date so as to complete the process within this time frame. The following information details the steps of the COA accreditation process from application through accreditation, maintenance, and reapplication and highlights differences, when they exist, for both new organizations and those seeking reaccreditation.

DETERMINING ELIGIBILITY AND APPLYING FOR ACCREDITATION

The first step in the accreditation process involves determining whether an organization is eligible. While an organization should conduct its own assessment, COA also determines eligibility for EAP accreditation at the time of application and bases its decision on a set of defined criteria that includes the following:

1. The organization must be an internal, external, or combined EAP and, at a minimum, provide the following core services:
   a. Employee Education and Outreach;
   b. Training to Supervisors, Managers, Human Resources, and Union Representatives;
   c. Management/Supervisory Consultation;
   d. Information and Referral, and Assessment and Referral Services; and
   e. Follow-up Referrals.
2. The EAP must have provided services to clients/employees for six (6) months at the time of application;
3. The EAP must hold all required governmental licenses/certifications for its services; and
4. The EAP must be sufficiently autonomous and independent to permit its review as a distinct legal entity.

After determining eligibility, an EAP interested in becoming accredited for the first time using the COA *EAP Standards and Self-Study*

Manual, 2nd Edition requests an application and submits information about the organization's services, finances, and operations.

Upon receipt of the application, COA sends the EAP a Financial Agreement that sets forth the accreditation fee. Once the signed Financial Agreement is received, COA sends the organization the *EAP Standards and Self-Study Manual, 2nd Edition* and instructs them to begin preparing their organization for accreditation.

INTAKE CALL

After the application phase, each EAP is assigned a designated COA Coordinator, an expert in the standards and process, who will assist the organization in becoming accredited. During an initial intake call, the Coordinator discusses the accreditation process and standards, matches an organization's programs to COA's service sections, and finalizes a service plan that includes a timeline for the self-study submission and site visit.

Coordinators typically conduct a full assessment of an organization's needs to determine what type of assistance they may need. When applicable, Coordinators may recommend training to organizations that need additional assistance. Training is another way that COA seeks to help organizations and prepare them for the process. COA conducts ongoing training to cover information regarding

- The content, format, and applicability of best practice standards;
- How to organize the self-study and prepare for the site visit;
- How to mobilize work groups and assign tasks with timelines;
- How to implement quality improvement mechanisms at all levels of the organization; and
- How to use standards to assess organizational readiness for accreditation.

Throughout the process, the accreditation coordinator and EAP contact person engage in frequent conversation to monitor progress and timely completion of the self-study document, to discuss and interpret standards, to discuss peer needs and plan for the site visit, and to prepare the presentation of any final materials to the Commission, COA's accreditation decision-making body.

SELF-STUDY

The self-study is a central component of COA's accreditation process and provides the first opportunity for an EAP to demonstrate compliance with the accreditation standards (see Appendix). The self-study is both a document and a process. As a document, it is a compilation of information from the organization (e.g., policies, procedures, minutes, training materials, etc.) that provides evidence of the EAP's compliance with the standards of best practice. As a process, the self-study is an opportunity for the EAP to engage in a course of self-assessment and work towards improving all operational functions including service delivery. The self-study process requires the participation and involvement of EAP staff, consumers, governing/advisory bodies, affiliates, and client companies, as applicable.

The self-study serves as the framework for the site visit process. Prior to the site visit, a Peer Review Team reviews information submitted with the self-study to determine compliance with the EAP accreditation standards.

Organizations typically take about six months to complete and turn in the self-study document to COA. COA expects an EAP to provide copies of its completed self-study to COA and to each of the members of the Peer Review Team at least eight weeks prior to the site visit so that there is adequate time to review the material.

While new organizations are required to submit pre-site documentation for all applicable standards, COA has recently redesigned the self-study phase of the process for organizations being reaccredited so that fewer pre-site documents need to be submitted. Instead of resubmitting documentation, the CEO will certify that certain documents have not substantially changed since the last accreditation review. Some of these organizational documents include Mission Statement, Bylaws, Grievance Procedures, Licenses, Certificates of Occupancy, Internal Accounting Procedures, Audit, and Insurance Policies. While the redesigned reaccreditation reduces the amount of pre-site evidence, organizations will still be responsible for the implementation of every applicable standard in the *EAP Standards and Self-Study Manual, 2nd Edition.*

SITE VISIT AND PEER REVIEW TEAM

The site visit follows the self-study and serves as an additional opportunity to provide documentation of an EAP's compliance with the accreditation standards and to allow the review and observation of the EAP's records, services, and facilities. Like the self-study phase of the process, the site visit is facilitative in nature meaning that Peer Reviewers will work with the organization to determine whether they are in fact meeting the true intent of a standard. While an organization receives their site visit report within forty-five days of the review, generally they already know how they've been rated because of the facilitative communication that occurs between the Peer Review Team and EAP during the site visit. The Peer Review Team measures compliance standard-by-standard. Pre-site documentation is reviewed prior to the site visit so that the team can get a general understanding of the organization's services and structure before conducting the on-site review.

A Peer Review Team is a group of two or more professional Peer Reviewers/Team Leaders, who meet COA's predetermined, written qualifications. Peer Reviewers and Team Leaders are individuals with extensive experience and expertise in the EAP field who undergo training and supervision with COA to understand the standards as well as the process of accreditation and their role in that process.

The Peer Review Team always consists of at least two Peer Reviewers, one of which acts as the Team Leader. The size of the team depends upon the size and scope of the EAP being reviewed. In selecting a Peer

Review Team, COA considers their background and expertise so as to complement the EAP's service mix and structure. In no case will a Peer Reviewer study an EAP when there is a conflict of interest. Peer Reviewers and EAPs are expected to notify COA of actual or apparent conflicts of interest, as soon as they are aware of their existence. COA reserves the right to make the final determination about Peer Reviewer assignment.

Once the Peer Review Team is finalized and approved, the Team Leader arranges a tentative site visit agenda, consulting directly with the EAP. The EAP is expected to accommodate all reasonable requests of the Team Leader.

Site visits span a minimum of one and half days. COA determines the site visit duration by considering the EAP's size, its services, and site and affiliate locations. COA reserves the right to extend the length of a site visit, if necessary to determine an EAP's compliance with COA's standards.

PRELIMINARY ACCREDITATION REPORT

Immediately following the site visit, the Peer Review Team prepares the Preliminary Accreditation Report (PAR) and mails the report to COA. The Peer Review Team determines the PAR ratings by using COA's EAP Weighting System and rating methodology. Once the report is received, a senior staff member at COA presents it in a twice-weekly PAR Committee Meeting. Each and every PAR is reviewed and reported on to ensure consistency in ratings, the accuracy of ratings and comments, and to make recommendations to organizations regarding how they can demonstrate compliance for those critical and mandatory standards that are still out of compliance. Within forty-five days of the site visit, COA sends the organization a copy of the PAR to allow further response before the Commission meeting. The EAP then takes the opportunity to respond in writing to the content of the PAR and submits a response to COA within forty-five business days.

ACCREDITATION COMMISSION

The Accreditation Commission is COA's volunteer decision-making body that reviews PARs and EAP responses for purposes of reaching accreditation decisions. Collectively, the Accreditation Commissioners are professionals with extensive backgrounds in the EAP field. The Accreditation Commission reviews all documentation in a manner free from conflict of interest and without knowing the identity of the EAPs under review.

An EAP approved by the Commission receives a four-year accreditation. Once an EAP is approved, COA sends a notification letter and Final Accreditation Report (FAR). The notification letter identifies the specific services for which the EAP has received accreditation as well as the expiration date. The FAR is a comprehensive management report listing the ratings for all standards as well as comments from the Peer Review Team/Commission for nonmandatory and noncritical standards that are still rated out of compliance. Organizations are expected to work on and bring into compliance prior to their reaccreditation all of the nonmandatory, noncritical standards listed on their FAR. The FAR also includes information from the Peer Review Team regarding the organization's strengths. In addition to the notification letter and FAR, the organization receives a plaque and other information to publicize their accreditation.

Occasionally, the Commission will make one of the following decisions: probation, suspension, denial, or revocation. A decision of probation, suspension, denial, or revocation typically arises when COA is made aware of reliable information that raises a concern about stakeholder health and safety, a serious risk management issue, and/or the credibility of COA's accreditation process. Probation and denial decision are typically granted to new organizations while suspension and revocation occurs to organizations seeking reaccreditation.

An EAP can also be denied accreditation for any of the following reasons:

- The EAP does not meet the eligibility requirements for COA accreditation at the time of decision making.

- The EAP submits self-study materials or information as part of the accreditation decision-making process that misrepresents the factual situation or is otherwise prepared dishonestly.
- The EAP fails to disclose information during the accreditation process that is or would have been germane to an accreditation decision.
- The unaccredited EAP holds itself out as accredited before formal notification by COA.

The Accreditation Commission also has the discretion to defer reaching an accreditation decision in order to allow the EAP to clarify its compliance with any standards with which the Accreditation Commission has questions. When the Commission defers a decision, they occasionally request that COA conduct a remedial site visit. More frequently, the organization is given between 3 and 9 months without a remedial site visit to correct any deficiencies and submit revised documentation showing implementation and compliance.

MAINTENANCE OF ACCREDITATION

During the four years that an EAP is accredited, COA requires the organization to maintain compliance with the EAP accreditation standards and to demonstrate continued compliance through completion of required reports, self-reporting of changes or events which could have an impact on continued compliance, and cooperation with any interim review processes, site visits, or external complaint review processes.

COA requires all accredited EAPs to complete a "Maintenance of Accreditation Report" annually. This document is a self-reporting device that apprises COA of incidents and occurrences as well as changes in services, structure, personnel, or funding or other factors that may raise questions about an EAP's continued ability to comply with COA standards.

REACCREDITATION

Eighteen months in advance of an EAP's accreditation expiration, COA notifies the EAP of their need to pursue reaccreditation. The reaccreditation of an EAP proceeds again from the agreement phase and all steps in the process remain the same except for the self-study phase.

CONCLUSION

Because of the rigor of the EAP standards as well as the length and detailed work of the process, COA recognizes that providing ongoing facilitation is a crucial element in helping organizations achieve and maintain best practices.

There is no doubt in the EAP field today that having common, industry-wide standards of best practice is important. Yet it is the process of accreditation under those standards that truly distinguishes one EAP from another. An EAP that chooses to become accredited and achieves this distinction is choosing to commit itself to a thorough, ongoing process of external examination and review as well as an internal process of organizational transformation and continuous quality improvement. This is the COA EAP accreditation process.

REFERENCES

The Council on Accreditation Policies and Procedures Manual, Council on Accreditation: New York, NY, October 2001.

The Employee Assistance Program Standards and Self-Study Manual, 2nd Edition. Council on Accreditation: New York, NY, January 2003.

CARF (The Rehabilitation Accreditation Commission). "Steps to Accreditation" [Online]. Available: http://www.carf.org/consumer.aspx?content=content/Accreditation/Steps.htm

Interlock Employee and Family Assistance Corporation of Canada: An Accreditation Case Study

Paula M. Cayley
Ulrike Scheuchl
Anne Bowen Walker

SUMMARY. The process of reaccreditation for Interlock was an extensive, often challenging, but ultimately exhilarating experience. It provided opportunities to grow as a company and led to the development of a number of new and improved systems and practices. This article describes Interlock's experience growing through this process. We attempt to define all of the strategic steps necessary for a successful achievement of accreditation. We also outline lessons learned in order to suggest useful guidelines for future applicants. *[Article copies available for a fee from The Haworth Document Delivery Service: 1-800-HAWORTH. E-mail address: <docdelivery@haworthpress.com> Website: <http://www.HaworthPress.com> © 2003 by The Haworth Press, Inc. All rights reserved.]*

Paula M. Cayley, MSW, RSW, is President and CEO, and an active contributor to the development of EAP standards; Ulrike Scheuchl, MS, is Project Manager; and Anne Bowen Walker, MSW, RSW, is Vice President of Clinical Services, Interlock Employee and Family Assistance Corporation of Canada, 4727 Hastings Street, Burnaby, BC, V5C 2K8 Canada.

The authors would like to thank Joan Deeks, PhD (cand.), for her professional support and contributions to this case.

[Haworth co-indexing entry note]: "Interlock Employee and Family Assistance Corporation of Canada: An Accreditation Case Study." Cayley, Paula M., Ulrike Scheuchl, and Anne Bowen Walker. Co-published simultaneously in *Employee Assistance Quarterly* (The Haworth Press, Inc.) Vol. 19, No. 1, 2003, pp. 45-60; and: *Accreditation of Employee Assistance Programs* (ed: R. Paul Maiden) The Haworth Press, Inc., 2003, pp. 45-60. Single or multiple copies of this article are available for a fee from The Haworth Document Delivery Service [1-800-HAWORTH, 9:00 a.m. - 5:00 p.m. (EST). E-mail address: docdelivery@haworthpress.com].

Digital Object Identifier: 10.1300/J022v19n01_04

KEYWORDS. Accreditation, employee assistance program (EAP) standards, Interlock, self-study manual, site visit

INTRODUCTION

Challenge, change and achievement marked the successful path of reaccreditation for Interlock. It was a multifaceted process that included analyzing established systems and practices, determining which systems would stay, change or be created, developing goals and timelines for completion of tasks, and finally doing the work needed to be accomplish this.

Whilst the reaccreditation process was stressful at times, the end result was extremly rewarding for the staff and organization alike. The lessons learned and the valuable experiences gained were worth it.

STRATEGIC STEPS TO SUCCESSFUL ACCREDITATION

The following is a detailed description of the strategic steps and the lessons learned in Interlock's successful accreditation experience.

Know Your Organization

"An employee assistance program (EAP) is a worksite-based program designed to assist: (1) work organizations in addressing productivity issues and (2) 'employee clients' in identifying and resolving personal concerns, including, but not limited to, health, marital, family, financial, alcohol, drug, legal, emotional, stress, or other personal issues that may affect job performance" (Employee Assistance Professionals Association [EAPA]).

Within this context, Interlock's purpose is to provide workplace results, assisting organizations to maximize their possibilities. To do this, Interlock conducts its business to be successful, supporting its capability and direction as a learning, growing organization.

To embrace the accreditation process was to ensure that the best practices of the EAP Industry were met. It is also a means to differentiate us from other EAPs and to demonstrate our intent and our capacity to provide exceptional service. Employee Assistance Accreditation Standards have been available since the 1980s and have been updated as recently as 2003. Going through the accreditation process would strengthen our commitment to enhance the quality of Employee Assistance services to customer organizations.

Interlock understood that the accreditation process would be a highly effective quality improvement initiative that would lead to the development of a number of new and improved systems and practices. It would also be a significant investment in the effectiveness and improved performance of the organization, contributing to overall resiliency.

INTERLOCK–THEN AND NOW

Interlock is an external Employee and Family Assistance Program (EFAP) provider founded in 1977. In the early 1980s it became a private not-for-profit organization subject to market discipline and led by a Board of Directors with nationwide representation of our customers.

The EAP movement began as a response to concerns about the abuse of alcohol and its effects on workers and the workplace. However, employers grew in their appreciation that a company's most valuable asset were its employees and that a wide variety of both personal and organizational problems could affect their productivity and, consequently, the productivity of the company. Throughout these developmental years of the EAP industry, Interlock's state-of-the-art services helped to define the scope of practice of EFAPs throughout British Columbia.

In 1997, Interlock became federally incorporated. With the globalization of the economy more employers in British Columbia had employees located throughout Canada.

At present, Interlock serves more than 350 organizations. The full range of EFAP services are available to all of the customer organizations.

All services are provided by Doctorate- and/or Master's-level, registered counsellors who are experienced EAP professionals.

In 1999, Interlock became the first British Columbia-based EAP to achieve accreditation through the Employee Assistance Society of North America (EASNA). In February 2002, Interlock applied for reaccreditation through the Council on Accreditation (COA). The accreditation process facilitated the redevelopment of a number of improved systems and practices, leading to the receipt of an expedited accreditation–the best possible outcome of this reaccreditation process.

ESTABLISHMENT OF AN ACCREDITATION COMMITTEE

Early in 2001, the CEO of Interlock presented the Board of Directors with a proposal that the company seek reaccreditation under the new accreditation standards. The CEO obtained the full commitment of the Board of Directors.

In August of 2001, the CEO assembled a team of key people to work on the project. It was cochaired by the Vice President of Clinical Services and a Regional Director of Interlock. In addition, external consultants assisted with various aspects of this project. They were

- a Human Resource Specialist that ensured that the Human Resource policies and procedures were brought into compliance with accreditation standards related to training and development of staff and
- a Program Management Coach that ensured that account management functions were consistent with the accreditation standards. This involved aligning policies and procedures for account management, updating contracts and establishing a new system for the effective renewal of contracts.

Employees were also invited to participate on the Accreditation Committee. All employees throughout the organization were committed to take the necessary steps for the accomplishment of the objective of COA accreditation.

To ensure a successful reaccreditation project, Accreditation Committee members were selected based on the following criteria.

Knowledge of the Accreditation Process

Interlock staff were engaged in a number of activities in preparation for accreditation. They were the following:

- The CEO completed a training session and subsequent peer-reviewer training at the EASNA conference in the spring of 2001.
- The Regional Director, who was part of the accreditation project when Interlock successfully obtained EASNA accreditation in 1999, was selected as he was key to the continuity of the process.
- The Vice President of Clinical Services attended the COA accreditation training session in Washington, DC, in September 2001, acquiring an overview of the accreditation process and an understanding of how the standards are structured, rated and weighted.
- In October 2001, Barbara Marsden, LISW, CEAP, President of EASNA, and Tim Stockert, MBA, MSW, Manager of EAP Services at COA, made a joint presentation on the accreditation process to the Accreditation Committee members and employees, providing an opportunity to clarify goals and objectives for this process.
- The CEO was a site review team leader for another organization, gaining invaluable knowledge about the preparation and process for peer review.

Skill Diversity

The CEO ensured skill diversity on the Accreditation Committee. There was a balance between people who were "process oriented" and people who were "results oriented." In addition, the team consisted of members with a variety of different roles and expertise within the company.

Team Orientation

Interlock knew that the key to a successful accreditation would be through teamwork. Steven R. Covey stresses in his widely acclaimed book *The 7 Habits of Highly Effective People* that "synergy means that the whole is greater than the sum of its parts." This effect was consistently demonstrated throughout the reaccreditation process.

The positive attitude of team members helped in times when the project seemed overwhelming. Everybody was committed to helping each other to ensure success.

The Accreditation Committee was fortunate to have people with a great sense of humor on the project to lighten the load when there was tension and stress. Introducing a little humor along the way contributed to employee morale.

The Board of Directors were kept well informed as to the progress of the project and continued in its commitment and support. The CEO continued to drive the project, monitoring progress and providing knowledge and support. The coleaders of the project had the knowledge and authority to be effective and had capable specialist and consultant assistance.

PROJECT PLANNING

The Interlock Accreditation Committee set up weekly meetings and identified responsibilities for each committee member. At the beginning of the accreditation process, the Accreditation Committee did not use a formal project management tool. However, during the process, the Accreditation Committee realized the need for a Gantt chart to manage the activities and timelines. Project planning involved

- identification of the tasks that needed to be completed;
- estimation of required time; and
- allocation of resources.

As tasks were assigned to Interlock employees, the progress of the work assigned needed to be monitored to avoid major delays and cost overruns.

The Gantt chart provided a tool for employees who were working on the project. The chart indicated the required tasks and steps to be taken on the vertical axis and the time scale across the horizontal axis. The Cochairs of the Accreditation Committee reviewed the employees' progress on a regular basis and provided assistance when necessary.

A key development, critical to the success of the project, was that in September 2002, the Accreditation Committee acknowledged the need for one committee member to work full time on the project. The CEO appointed a Project Manager, whose responsibilities were to track assignments, monitor work completion, and edit and assemble the Self-Study Manual.

A process of distributing questionnaires to stakeholders, members of the Board of Directors, clients, employees and affiliates was set up as part of project planning. Questionnaires were returned directly to COA. A list of the recipients was submitted to COA as part of the Self-Study Manual. Interlock followed up with recipients to encourage completion and return of the questionnaires.

During the preparation of the self-study document, the Accreditation Committee assessed the progress and identified the following major tasks that still needed to be completed.

Creation of a Revised Policy and Procedures Manual

The policies and procedures needed to be redesigned to reflect both Interlock's vision and the EAP accreditation standards. The revised manual incorporated the newly drafted policies and procedures, a new organizational structure.

To ensure a consistency throughout the manual, three members of the Accreditation Committee reviewed and edited the policies, procedures and protocols. Key to the process was the close attention to detail throughout the manual.

The outcome of the process resulted in a more user friendly, professional manual which included policies, procedures and protocols in the following areas: Operations and Administration, Health and Safety, Quality Improvement, Human Resources, Contract and Account Management, and Clinical Services.

The policies and procedures were reviewed and approved by the CEO and then communicated to all employees in company-wide training sessions. Affiliates were sent copies of the applicable policies and procedures and were invited to participate in training sessions.

Updating Client Record Forms

The COA standards raised requirements for maintenance and content of client records. All new requirements needed to be assembled and incorporated in the new Client Record booklet. In addition, the Accreditation Committee took the opportunity to track other specific concerns for employers such as depression in the workplace. The result was a highly functional document, professional in appearance and both clinically and legally sound.

Development of a Quality Improvement Planning System

As part of the quality improvement planning process, Interlock formed a Quality Improvement Committee which reported to the Board of Directors. It consisted of the CEO of Interlock, the Vice President of Business Development and the Vice President of Clinical Services. The Quality Improvement Committee initiated systems to reposition Interlock for ongoing improvement in performance and growth.

A formal Organization Satisfaction Survey was also created to provide information that allowed Interlock to partner more effectively with customer organizations. In addition to recipients responding to a number of questions in the Organization Satisfaction Survey, respondents were also given the choice of receiving a phone call or scheduling a meeting with the CEO to further discuss service delivery issues.

A Quality Assurance Report was initiated and is now completed on a quarterly basis by all Regional Directors. The report includes data on risk reports, critical incident reports, referral resources, client record review, clinical supervision, case consultation, and other analyses.

Annually, company-wide results of Interlock's quality assurance and evaluation are combined with the goals and objectives of the strategic plan to produce a written annual Quality Improvement Plan. The overall objective of this annual plan is to improve the quality of services provided to clients and organizations and to continue to improve the quality services provided by Interlock.

Interlock's Quality Improvement Plan is a cyclical four-part process:

- A quality improvement issue is identified and reviewed to assess the success of a plan and make further adjustments if necessary or to correct a problem or implement a change initiative.
- Once the issue is identified with all feedback and relevant data reviewed, the Quality Improvement Committee or designates would meet to develop a solution to the identified problem and/or plan for change.
- The plan or directive is implemented. This plan or directive could involve a process, a form, a product or a department.
- Feedback through surveys, word of mouth, data monitoring, observation or other means is collected, tabulated and summarized to determine the success of the issue's outcome. This evaluation can lead to additional action plans.

INVOLVEMENT AT ALL LEVELS–PARTNERSHIP IN ACTION

It is important to individuals throughout the organization in the accreditation process. The Board of Directors and the Senior Management Team had a firm desire to see improvements at Interlock and strongly supported the accreditation initiative.

Ongoing communication and training initiatives were key to keeping employees informed about the accreditation process. This was accomplished in several ways:

- In October 2001, a presentation on the accreditation process was given by Barbara Marsden, President of EASNA, and Tim Stockert from COA.
- To meet accreditation standards related to training and development of employees, Interlock redeveloped a comprehensive in-house training program tailored to the needs of the organization. These training sessions were held on a quarterly basis with all employees throughout the company.
- Training packages were assembled and distributed to the affiliates. Interlock also ensured affiliate competence in EAP practices through training sessions.
- To keep employees informed about the project, memoranda and voice mail messages were regularly distributed.

The accreditation project required change in Interlock operations. Acceptance of the business operation changes by the employees would be critical, in order for Interlock to move forward in the accreditation process. Employees also needed to understand that the accreditation project may affect their duties and responsibilities. For example, job descriptions had to be updated to match accreditation, that is, specifically detailed to demonstrate to customers and potential customers Interlock's adherence to best practices.

In addition to internal communication, the Accreditation Committee maintained regular contact with the COA. Tim Stockert was very helpful in responding to questions at different stages along the accreditation process. It was also very helpful to talk to other employers who had successfully completed the COA accreditation process. There were numerous people who mentored the Accreditation Committee along the way. As a result, Interlock would strongly recommend the EASNA mentoring program for EAP undergoing COA accreditation.

SELF-STUDY MANUAL AND GUIDELINES

A central component of the COA accreditation process was the self-evaluation. Using the COA's EAP standards for preparation of the required materials for the self-review process, it became evident that changes needed to be made to some of Interlock's practices and procedures. The reaccreditation project required input from employees. All employees were involved in the process either directly or indirectly. Direct involvement consisted of those employees who were active in rewriting policies and procedures and indirect involvement, as the new practices and new procedures required acceptance, openness and responsiveness from all employees.

Step 1: Specific material necessary to demonstrate compliance with each standard was provided to the Project Manager, who organized the pre-site documentary evidence, including forms, policies and procedures.

Step 2: The Project Manager assembled the Self-Study Manual. It was prepared according to the guidelines provided by COA.

The Self-Study Manual consisted of seven 3-inch binders. Detailed attention was paid to all mandatory standards to ensure full or substantial compliance. Based on this review, some of the responses to the standards needed to be revised. At the same time, the Project Manager checked the manual for completeness.

Step 3: In February 2003, Interlock sent a copy of its Self-Study Manual to COA. Copies of the Self-Study materials were sent to the Peer Reviewers in March 2003.

PREPARATION FOR THE SITE VISIT

Since Site Reviewers have a limited amount of time during the site visit, materials needed to be readily accessible at all times. All required documents were maintained in a room that was reserved for reviewers throughout the site visit.

To prepare for the Site Visit, the Accreditation Committee gathered additional documentation referred to as on-site evidence in an effort to demonstrate compliance with the standards. The materials included a copy of the completed Self-Study Manual, all applicable licences, contracts with customer organizations, personnel manual and employment records, job descriptions, policy and procedures manual, training materials, board of directors manual, meeting minutes, client records and other documentation to demonstrate compliance with accreditation standards.

SITE VISIT

The Site Visit was an opportunity for Interlock to demonstrate full compliance with the COA accreditation standards. The Peer Review Team from COA consisted of two highly skilled professionals who had extensive experience in the EAP industry.

The Site Visit took place over a three-day period. During these three days, the Peer Review Team assessed Interlock's systems and processes.

The site visit entrance meeting was the first order of business, and it was attended by the Chair of the Board of Directors, the management team and employees. Following the site entrance meeting, a selection of Board of Directors, employees, affiliates and customer organizations were interviewed in an effort to determine Interlock's ability to demonstrate best practices in the EAP industry. Interlock's accountants and lawyers were also interviewed as a means of assessing our strength in complying with Canadian authority and legal standards. Interlock Head Office, centralized services department and two of the Regional Offices were visited and thoroughly scrutinized.

During the exit interview by the Peer Reviewers, Interlock was given some general feedback about the exceptionally high quality of the Self-Study Manual. In addition, Interlock was advised that it had demonstrated a clear commitment to best practices.

POST-ACCREDITATION SITE VISIT

A debriefing at the end of the accreditation site visit was important to Interlock stakeholders. Debriefing meetings were set up with three different groups:

- CEO, Accreditation Committee Cochairs, Project Manager;
- Interlock's Leadership Team; and
- CEO, Project Manager, external Consultants

These meetings provided the opportunity to take a more objective retrospective analysis of the accreditation processes. We reflected on the necessity of accreditation, the excitement of the process and the complexity of the project. At the same time, there were many challenges to achieve the goal of accreditation. The official documentation confirming Interlock's accreditation was received approximately 2 months after the site visit.

ELEMENTS OF A SUCCESSFUL ACCREDITATION AT INTERLOCK

Company-Wide Understanding of the Accreditation Process

It was essential that there was a company-wide understanding of why Interlock was committed to this process. Everyone needed to see how the process and the outcomes fit into Interlock's goal of being the best that we can be. Being accredited under the COA standards was a way to differentiate Interlock and demonstrate value to our clients, our business customers, potential customers and to the EAP industry as a whole. An understanding of this purpose kept employees focused on the overall goal of Interlock, as well as a greater appreciation of the accreditation and its relevance to our organization. Again this required a dedication to ongoing communication with everyone associated with this project.

Assignment of a Project Manager at the Early Stage of the Project

Another lesson learned was that a project of this magnitude requires a dedicated Project Manager from the early stage of the process and not someone with many competing tasks and priorities. The Project Manager would have the required authority, either given or implicit, to coordinate the progress of the project within the required timeframe. The Project Manager would also need strong administrative, technical and organizational skills.

Coordination of Drafting of Policies and Procedures

Interlock involved numerous employees in the writing of responses and gathering of evidence, each responsible for one distinct section. Although this proved to be a great learning experience transforming more people's awareness of the purpose and magnitude of the project, it was not effective in producing results. Interlock realized that it was more beneficial to have someone immersed in the process to truly grasp the essence of what was required. The wide range of personalities, thinking and writing styles did not lend itself well to achieving the desired goals in a time-sensitive, cost-effective manner.

Awareness of the Magnitude of the Project

It is crucial that the company is aware of the magnitude of the accreditation and that it is more than just a written report on what exists within the company. Accreditation is a full commitment to review and where necessary to rebuild the internal structures of an organization, the practices, the policies, and the tracking systems. At Interlock, the Accreditation Committee redesigned the Client Record booklet and the Board of Directors' Manual as well as revising the Policy and Procedures Manual and numerous other documents. These revisions helped bring Interlock into compliance with the new standards. The scope of the project provided the leverage for many changes to past systems and procedures.

Review of the Final Draft of the Self-Study Manual by an Internal Person Who Was Not Directly Involved with the Drafting of the Manual

Prior to submitting the completed Self-Study Manual it was reviewed by the CEO who, until that point, had remained essentially uninvolved in the details of the Self-Study Manual. This allowed for a review by someone who had tremendous knowledge of Interlock and the EAP industry but who had not been directly involved in the development of the self-study.

The Need for a Functionally Diversified Accreditation Committee

Another of Interlock's realizations was the need for a functionally diverse accreditation team. The team not only includes a range of expertise but team members have to appreciate each other's differences and contributions and be able to work well together, especially under pressure.

Seek Assistance Early from EASNA'S Mentoring Program

Interlock would recommend that any organization embarking on accreditation prepare itself in advance with as much knowledge as possible about the process. Using the EASNA mentoring program and benefiting from the experiences of those who have completed the process would be advantageous. We would also recommend that COA provide a more user friendly and more detailed study guide and template.

CONCLUSION

Although reviewing notification of the COA accreditation may have been the final step in the accreditation process, it is really just the beginning. While accreditation was achieved Interlock's philosophy of continuous growth will keep the accreditation in the forefront of the organization and will remain a work in progress with the aim on future reaccreditation.

REFERENCES

Council on Accreditation (2003). *Standards and Self-Study Manual Employee Assistance Programs. 2nd edition.*

Council on Accreditation Training Institute (January 2001). Preparing for Accreditation for Larger, Smaller, Internal and External EAPs.

Covey, S. R. (1990). *The 7 Habits of Highly Effective People.* Simon & Schuster Inc.

Drotos, J. C. (1999). EAP Accreditation and Credentialing. In *The Employee Assistance Handbook,* edited by James M. Oher. John Wiley & Sons, Inc.

Employee Assistance Professionals Association (EAPA), http://www.eapassn.org/public/pages/index.cfm?pageid=507

Kerzner, H. (2003). *Project Management: A Systems Approach to Planning, Scheduling, and Controlling.* John Wiley & Sons, Inc.

Magellan Behavioral Health: A COA Accreditation Case Study

Christina H. Thompson

SUMMARY. Magellan Behavioral Health successfully earned the 4-year Council on Accreditation (COA) accreditation that is supported by the development of new Employee Assistance Program (EAP) standards. Magellan's experience is outlined with discussion of various components, experience and lessons learned while going through the process. *[Article copies available for a fee from The Haworth Document Delivery Service: 1-800-HAWORTH. E-mail address: <docdelivery@haworthpress.com> Website: <http://www.HaworthPress.com> © 2003 by The Haworth Press, Inc. All rights reserved.]*

KEYWORDS. Council on Accreditation, employee assistance, Magellan, EAP standards of accreditation

Christina H. Thompson, LCSW, MSWAC, CEAP, is Vice President of Employee Assistance Programs and Addictions Services, Magellan Behavioral Health Services, 6950 Columbia Gateway Drive, Columbia, MD 21046 (E-mail: ththompson@ magellanhealth.com). Mrs. Thompson has been in the behavioral health field for over 23 years.

[Haworth co-indexing entry note]: "Magellan Behavioral Health: A COA Accreditation Case Study." Thompson, Christina H. Co-published simultaneously in *Employee Assistance Quarterly* (The Haworth Press, Inc.) Vol. 19, No. 1, 2003, pp. 61-71; and: *Accreditation of Employee Assistance Programs* (ed: R. Paul Maiden) The Haworth Press, Inc., 2003, pp. 61-71. Single or multiple copies of this article are available for a fee from The Haworth Document Delivery Service [1-800-HAWORTH, 9:00 a.m. - 5:00 p.m. (EST). E-mail address: docdelivery@haworthpress.com].

THE BETA TEST

In 2000, Magellan was invited to participate in the Council on Accreditation's (COA) beta testing of its newly developed Employee Assistance Program (EAP) standards. Magellan's senior leadership felt that as the nation's leading provider of EAP services, this was a unique opportunity to reinforce our commitment to quality improvement.

Due to the considerable amount of EAP stand-alone business conducted at Magellan's Salt Lake City national call center, the company selected that particular office as the test site. Although Magellan regularly conducts internal audits, customer reviews, and participates in other lines of behavioral health accreditations, this was the first such accreditation experience for our EAP business.

The beta testing involved several activities and was designed to evaluate and refine the standards, timelines, process, self-study, and site visit guidelines.

Magellan's involvement in the beta testing allowed the company to experience firsthand the efforts required both pre- and post-review, while at the same time allowing us the opportunity to gain hands-on experience and a better understanding of the time commitment, resources, infrastructure, and expense necessary to achieve and maintain EAP accreditation.

OVERVIEW

Immediately after the first version of the COA standards was released in 2001, Magellan began preparing for the Accreditation. The core steps of the process included the following:

- Identifying participating locations.
- Distributing the standards and self-study manual to all the sites.
- Identifying those resources that would need to be committed to the effort.

- Communicating with other Magellan offices and departments about their role and how the effort would impact all areas of the company.
- Creating a detailed project plan organized by assignment, timeline, and assigned point of contact.
- Drafting the self-study.
- Preparing for the site visit.
- Conducting the actual site visit.
- The maintenance involved once accreditation was achieved.

It should be noted that the aggressive timeline required that a number of these core steps happen simultaneously.

SITE SELECTION AND LOCAL STAFF PARTICIPATION

Magellan's EAPs serve over 15 million members throughout the United States and in 26 other countries with business run out of several call centers. The focus of the accreditation was on EAP stand-alone business, so it was necessary to include call centers that managed the majority of our stand-alone EAP business. These sites included Alaska, California, Missouri, Utah, Washington, along with various staff offices. Unless contractually specified, our processes are standardized. This was demonstrated at the various site visits. Once the sites were selected, it was time for the real work to begin.

We decided to have our corporate offices, located in Maryland, handle several sections of the self-study and materials, enabling the local offices to have staff participation around the following key functions: clinical, quality improvement, operations, network, systems, training, account management, and human resources.

DISTRIBUTION OF STANDARDS

With COA's permission, we distributed the standards to the selected sites for review. Weekly meetings were set up to discuss the various aspects of the process, share best practices, and provide updates. Each site was assigned a project manager.

The first meeting outlined the activities and delineated those tasks that would be handled by corporate and those that would be coordinated at the sites. The initial focus of these meetings was on the pre-site or the self-study activities. COA recommends at least six months to prepare for the self-study, but our timing afforded us nearly eight. The standards had undergone a number of changes since the initial beta test, and given that and the number of people involved at the various sites, we welcomed having the extra time for this considerable undertaking.

CONTINUED ORGANIZATIONAL SUPPORT

Communication and education to all levels of staff within the organization about the processes and impact was critical to our success. Updates by the cochairs were provided both in standing meetings and in writing to the senior leadership team on a regular basis. We also kept detailed records on the amount of time spent on each activity. The activity time was then converted to actual costs and combined with the costs of materials in an attempt to quantify the administrative expenses in addition to the fees charged by COA.

Senior leadership continually evaluated the value that this process would bring to the company, the industry, our customers, members and network providers. One of the more significant challenges involved ensuring that the process didn't interfere with the day-to-day functions and responsibilities of these sites. Fundamentally, however, we were confident that the accreditation would serve the interest of our stakeholders and become another market differentiator for the company.

THE PROJECT PLAN

Not surprisingly, given the complexity of the process, our Accreditation Project Plan served as a road map for our success. The time and energy into developing and maintaining this kind of detailed account was worth the effort. Once the initial plan was completed, we updated it weekly and referred to it often as both a resource and guide.

The project plan was broken out into key categories:

- Administration and management
- Health and safety
- Finance
- EAP legal liability
- Controls
- Quality Improvement initiatives and activities
- Personal and affiliate competence
- Staff supervision
- Professional practice
- Access to services and referrals
- Service delivery

All of the standards for each key category were further broken out within each standard, to include self-study and on-site requirements. For each standard, the plan listed where the evidence would

- be documented;
- need general comments or notes;
- need to be completed and a due date for that completion;
- show the results of our own internal audit of the materials;
- include someone to be interviewed/observed as evidence;
- be a pre-site requirement;
- be an on-site requirement; and
- have a date it was last updated or the version in place.

Each of the above also identified a Magellan team member whose responsibility it was to complete the related task, the person to whom the information needed to be submitted and by when, and whether or not there were any required client account customization of which COA would need to be made aware.

We also felt it was important for our staff to have a complete under-standing of what the reviewers would be looking for and their perspec-tive, so several of our internal accreditation team attended the COA peer-reviewer training.

Drafting the Self-Study

In our experience, it was this portion of the process that required the most preparation time. There were several important elements that we would recommend. Some of which include the following:

- Develop a system for collecting the various section materials. In our experience, we found that one person worked best, but cer-tainly no more than two.
- Make sure the point people clearly understand what tasks they are responsible for, to whom it should be delivered, and the deadline. As the chair, I was selected as this primary point person. I was re-sponsible for seeing that all the sections were complete, compiled, and sent in based on the requirements detailed in the standards, as clearly laid out in the self-study. It was also my role to identify any deficiencies in the documents and decide how best to address those issues.
- Be vigilant about making your deadlines. Since we were coordi-nating multiple locations, we found this discipline to be invalu-able.
- Early in the process, be very clear about what documentation is re-quired. This will help focus the scope of preparation and limit the need for last minute searches, rewrites, and time spent collecting unnecessary material.
- If you find that policies, workflows or other documents need to be revised, updated, or require staff training, plan for this as early as possible.
- Use COA as a resource. Open dialogue with a point person at COA was helpful in making sure that we interpreted the standards cor-rectly. We had a number of questions about specific terms, mea-

surements, and intent of some of the standards. If there is the slightest question, be sure to ask.

- Make sure you have adequate resources to support all the self-study components, as this is a very labor-intensive process. Be aware that the process also will require additional expenditure of funds to cover preparation, implementation, and monitoring.

- Once all the materials are compiled, make sure to carry out one last quality check. This will help not only in terms of the professional appearance of your materials but also to ensure that all of COA's requirements are met.

- Make sure the materials are easy to follow. Do not assume that your reviewers are familiar with your particular program. Your materials must be clear and concise. Done properly, this effort can save time later on during the on-site activities, since fewer questions or requests for additional materials may occur.

- Carefully follow your progress. For example, there are survey requirements that need to be completed early in the process to ensure that COA has adequate time to review the results prior to the site visit. Since we have our own contractual requirements for distributing and evaluating satisfaction surveys, this proved to be a challenge.

PREPARING FOR THE SITE VISIT

It was critical that each of our selected sites were fully prepared for their site visits. Each site received a copy of the self-study and learned how the focus was supported by the project plan for the on-site requirements. This involved the following:

- Compiling the additional materials that would be available to the reviewers while on site. Although some materials were not required as evidence in the self-study, they might be considered proprietary and would need to be addressed one-on-one with Magellan staff. For example, it was necessary to have human resource staff available at the time of the visit to review personnel records.

- It was very important to make sure that our staff was adequately prepared and fully understood the various requirements. This was addressed during routine unit staff meetings at all levels, and included quizzing staff and conducting mock interviews. Our staff later reported that these elements helped them feel more confident during the actual site visits and interviews. One site developed games to reinforce key sections by using flash cards, word searches, and contests to include an element of fun in the process.

- We developed a PowerPoint presentation that gave a broad overview of our business and processes on day one of the site visit to the reviewers and invited each site to customize it with the unique elements of their sites. This allowed the reviewers all to have the same basic overview of Magellan's EAP business and set the tone for what they would be seeing over the next two days of the visit.

- An element of the process was the selection of reviewers. We were able to work with COA to strike a balance between the right expertise and, quite frankly, concerns about having competitors on site. We also found that it was helpful to understand the reviewer's unique area of expertise. For example, an internal program may have very different processes than an external one, or a small, local program may have a very different infrastructure than a large, external program. It is helpful to try to anticipate what components may need further clarification to make sure the reviewers appreciate the nuances that may be different than their own experiences.

- A simple yet important task was making sure that the reviewers had a separate space to work that not only afforded them privacy but also did not disrupt the day-to-day operations of the site or compromise confidentiality. We also created a directory of local restaurants, hotels, and other amenities to help make the reviewer's stay more pleasant.

- Our cochairs visited the sites two or three weeks prior to the actual review to do a dry run. Materials were checked, staff interviewed, random charts audited, observations done, and a checklist was created and completed for readiness.

THE SITE VISIT

This was one of the easiest parts of the process. Our careful preparation helped us feel confident and ready for a positive outcome. We developed a detailed agenda and timeline of the activities that would occur over the next two days with the reviewers. We did not know which reviewer would perform which tasks, so we revised the agenda accordingly. We found that flexibility and the ability to make last-minute changes to be important.

At the end of the two days, the reviewers gave the group basic impressions at a wrap-up meeting, noting that the final report would include greater detail and, of course, the final scoring. Our committee could ask questions, knowing that the reviewers were limited in what they could say.

When the visit was complete, our staff was relieved, encouraged by the input we had received, but still apprehensive. Had we passed or not?

THE RESULTS PERIOD

It takes about 45 days to receive the final report. Although we had received mostly threes and fours in the initial scoring, there were also opportunities for improvement in areas outside of Magellan's internal processes, primarily related to our affiliate network. Taking that feedback into consideration, we adjusted our related processes to better reflect and support the standards and to make sure these areas were readily identifiable and evident in the chart documentation. We revised a number of forms and conducted trainings and outreach to the affiliates with respect to the accreditation expectations. We had six months to implement these new processes and took this time to make a number of improvements. COA also had a few other requests for clarification or additional information, and we were able to provide that to them.

We were very proud when we received word that we had passed and were accredited for four years for every level offered by COA's EAP

Accreditation, and were particularly proud to be one of the few initial programs to be dually accredited by both COA and EASNA for the four-year time frame. We opted to go for all the various levels offered by the accreditation but note that all are not required to go through the process, since all programs may not offer some levels. The services covered for us include the following:

> Core EAP Services: Employee education and outreach; Critical Incident Stress Management; training to supervisors, managers; human resources; and union representatives; management/supervisory consultation; information and referral; assessment and referral services and follow-up referrals.
> Additional EAP Services: Short-term counseling/short-term problem resolution; organizational development; drug-free workplace services; work-life services and legal services.
> Modalities of Service: Services via online and telephone modalities; services via Website and international services.

MAINTAINING ACCREDITATION COMPLIANCE

Successfully achieving accreditation is a wonderful testimony to the expertise and professionalism of our organization. However, it is not the end of the process. Maintaining the standards of accreditation on a day-to-day basis is critical. We continually evaluate our processes to make sure that we are meeting or exceeding the standards. Since most of what we do is standardized, our integrated business units are also using the same processes as "best practices," even though COA is not yet looking at integrated options.

We maintain a schedule to make sure documents are reviewed, updated, and trained on to comply with the accreditation expectations. Any change in our processes or organization that would impact what the reviewers observed is communicated with COA in writing immediately to make sure that we remain in compliance. Maintaining accreditation requires infrastructure support and does factor in the additional expenses. We continue to market our achievement and hope that more organizations will realize its value not only to their companies but also to their customers and to the EAP industry as a whole. It definitely strengthened our internal processes and enabled us to continually monitor the quality services we offer.

Issues in International Employee Assistance Program Accreditation

Dale A. Masi

SUMMARY. This article describes the present size of Employee Assistance Programs outside of the United States. It emphasizes the rapid growth and the development of the profession beyond the Employee Assistance Professionals Association (EAPA). It then discusses the international programs and approaches to accreditation. Suggestions are made for international agencies that might be apprehensive about standards being too "American" and not being cross-culturally applicable. The Council on Accreditation (COA) appears the most logical, but the author also describes the recent development of worldwide guidelines, which have been sponsored by numbers of EAP groups. The guidelines might be a pathway or intermediary step to accreditation for those in-

Dale A. Masi, DSW, CEAP, LICSW, is Professor, University of Maryland Graduate School of Social Work, and Director, Employee Assistance Program specialization at the University, and President/CEO, Masi Research Consultants, Inc. She also Chairs the U.S. Government Sponsored Joint Industry Public/Private Employee Assistance Program Alliance.

Address correspondence to: Dale A. Masi, 2549 Virginia Avenue NW, Washington, DC 20037 (E-mail: <masirsrch@aol.com>; <www.eapmasi.com>).

[Haworth co-indexing entry note]: "Issues in International Employee Assistance Program Accreditation." Masi, Dale A. Co-published simultaneously in *Employee Assistance Quarterly* (The Haworth Press, Inc.) Vol. 19, No. 1, 2003, pp. 73-85; and: *Accreditation of Employee Assistance Programs* (ed: R. Paul Maiden) The Haworth Press, Inc., 2003, pp. 73-85. Single or multiple copies of this article are available for a fee from The Haworth Document Delivery Service [1-800-HAWORTH, 9:00 a.m. - 5:00 p.m. (EST). E-mail address: docdelivery@haworthpress.com].

ternational EAPs that do not feel they are prepared to undergo the formal COA process. *[Article copies available for a fee from The Haworth Document Delivery Service: 1-800-HAWORTH. E-mail address: <docdelivery@ haworthpress.com> Website: <http://www.HaworthPress.com> © 2003 by The Haworth Press, Inc. All rights reserved.]*

KEYWORDS. EAP international demographics, international EAP programs, worldwide guidelines

INTRODUCTION

At present, there are 4,412 EAPA members in the United States. Of these 518 (13%) are international members not residing in the United States (13% of total membership). There are 3,337 individuals in the United States that are CEAPs and 83 from other countries.[1]

This does not, however, in any way provide us with a comprehensive picture of the size of the EAPs in other countries than the U.S. For example, within the past 2 years, a group from Europe called the European Forum composed of EAP practitioners from the United Kingdom to Russia was formed. A similar group was also convened in Asia 2 years ago, and more recently (October 2003), the first meeting of the Central and South American EAPs convened. All of these three are separate entities from EAPA, and most members are independent of EAPA.

The question of accreditation is critical for the development of the EAP program internationally. Already, international programs are applying for accreditation by the Council on Accreditation (COA) in New York. This includes EAPs from the United Kingdom, Puerto Rico and Japan. As the author collaborates with the latter two, it is clear that cultural differences have to be taken into account.

Programs from other countries applying for accreditation should first review the self-study manual (obtained from COA) and list whatever appears to be a problem for them, or what they do not understand. They also should list what documents they do not have in English. The author is recommending a translator from the native country be engaged for the self-study.

There has been a virtual explosion of EAP programs worldwide. In Geneva, Switzerland, in the fall of 2002, a meeting was convened of EAP organizations worldwide, including

- Council on Accreditation (COA)
- Employee Assistance Professionals Associations (EAPA US)
- Employee Assistance Professionals Associations International (EAPA Int'l)
- Employee Assistance Professional Associations of Australia (EAPA-A)
- Employee Assistance Society of North America (EASNA)
- European Network on Occupational Social Work (ENOS)
- International Council on Alcohol and Addictions (ICAA)
- International Federation of Social Workers (IFSW)

At the same time, there was great concern about the lack of uniform standards of practice for the EAP field. As a result, the author was commissioned by the organizations below to develop worldwide guidelines for EAPs in the workplace. Documents were provided by the following organizations:

- Employee Assistance Professionals Association–International Program Guidelines.
- EAP National Guidelines–Australia/New Zealand (the most recent version of EAPA was received after the guidelines were drafted).
- Employee Assistance Professionals Associations European Forum.
- Article of Associations of the European Network on Occupation Social Work (ENOS), 1995.
- Conceptual Framework for the Professional Field of Occupational Social Work, Federal Association of Occupational Social Work ENOS–Germany.
- Global Qualifying Standards for Social Work and Education and Training, International Federation of Social Work (IFSW) Bern, Switzerland.

- Government Guidelines for Workplace Mental Wellness Health Care System Tokyo, Japan.
- International Council of Alcoholism and Addictions, 1999.
- Management of Alcohol and Drug-related Issues in the Workplace–International Labor Organization, 1996.
- Draft Code of Practice on Managing Disability in the Workplace–International Labor Organization, 2001.
- New Standards Guidelines–COA.
- UK Guidelines for Audit and Evaluation for Employee Assistance Programs UK EAPA.

The author then developed a matrix where she was able to extract common principles for international procedure.[2] They are listed below.

MISSION/OBJECTIVES

It is essential that an EAP have a clear mission statement.

ETHICS/VALUES

The EAP adopts and follows its own code of ethics, requires its professional staff and affiliates to adhere to the codes of ethics of their respective professions and avoids conflicts of interest in carrying out its responsibility.

CONFIDENTIALITY

The EAP gives clients written information that describes the EAP's confidentiality policy, and requires clients to sign a statement indicating their understanding of confidentiality rights and limitations.

The EAP follows policies and procedures governing access to, use of, and release or disclosure of information about clients and such policies meet applicable legal requirements under federal, state, or provincial law.

RIGHTS OF EMPLOYERS

Employers should provide and maintain a safe and healthy workplace in accordance with the applicable law and regulations, and take appropriate actions.

Employers should have the right to take appropriate measures with respect to workers with alcohol- and drug-related problems which affect, or which could reasonably be expected to affect, their work performance.

CLIENT RIGHTS

In recent years, organizations representing people with disabilities have worked to make disability an issue of equal rights, rather than social welfare, and to generate change in opportunities for participation in employment and society. They have effectively challenged what is termed the "medical" model of disability, which focuses on the individual's impairment and inability to perform certain everyday activities, including work activities, and views solutions in the context of individualized rehabilitation programs. In its place, they have promoted the "social" model of disability, in which the focus is on the constraints arising from social, political, economic and cultural factors, as well as barriers in the built environment, and on solutions through measures to remove these constraints and barriers.

MANAGEMENT/BOARD OR ADVISORY GROUP

An EAP if not governed by a Board of Directors should consider having an Advisory Board.

EAP POLICIES

There are two policies to consider in an EAP–A policy for the program and an EAP policy for the EAP organizations.

EAPA expects that its members will ensure that, wherever possible, an organization operating or implementing an EAP shall have a policy, which defines the purpose and objectives of the service and its relationship with other organizational functions. The program shall be clear and readily accessible to everyone involved.

PROGRAM DESIGN

Create a Workplace Mental Wellness Policy, which incorporates Mental Wellness Systems at the workplace, assessment of workplace problems and implementation of mental health care, and professional personnel, which provide mental health care.

IMPLEMENTATION

An implementation plan outlines the actions needed to establish a fully functioning EAP and sets down a time frame for completion.

SUPERVISORY MANAGEMENT CONSULTATION

The EAP staff consults with key staff regarding the management and referral to the EAP of employees with job performance problems.

RECRUITMENT, TRAINING AND STAFFING

The program should have a formula for adequate staffing. The EAP should be staffed by an adequate number of individuals specifically trained in Employee Assistance Programming (program managers as well as counselors).

UNION REPRESENTATIVES

It is important for the success of EAPs that management/union representative proactively support the introduction of EAPs into organizations, help develop policy and procedures, provide resources to operate, and encourage employees to use the EAP when necessary.

SERVICES DELIVERED

Program Promotion, Education

Booklets, brochures and posters and other promotional material should be readily available and accessible to all employees during the life of the program. EAP training is developed in such a way that it can be incorporated into an organization's existing procedures and training schedules. All supervisors and mangers should participate in workshops in EAP referral procedures and practices. All employees should be made aware of an EAP through an organization wide awareness campaign.

Prevention Services

The EAP provides prevention services that address the following components:

- outreach;
- health promotion and wellness; and
- coordination with health care providers

Topics addressed in prevention activities are changes and updated to reflect the needs and feedback of the host or customer organization and its employees.

The EAP emphasizes the importance of prevention in all of its activities and offers to provide at least one primary prevention activity annually, for the host or customer organizations.

The EAP develops and offers educational sessions on wellness and other prevention-related topics.

Critical Incident Stress Management

EAPA standards expects its members to ensure that the purchasing organization has given careful thought as to how employees are supported in urgent, serious or emergency situations in a timely fashion and consistent with organizational policies. Wherever possible this service should be provided by the EAP.

Access Procedures

Procedures for accessing EAP services that minimizes barriers to the timely initiation of services or use of services and give priority to employees or eligible participants with urgent needs or in emergency situations are important.

The EAP communicates to customer employees, and eligible participants that access to the EAPs services occurs through one of the following mechanism:

- self-referral by employees and eligible participants for problems that may be adversely affecting their job performance;
- referrals by supervisors and suggestions by union representatives, human recourses, and/or medical personnel; and
- mandatory referrals.

Work Life

The EAP conducts a needs assessment of the host or customer organization to determine the most appropriate and effective work-life services for the host or customer organization and its employees.

EAP assessment procedures include the use of a work-life intake tool to evaluate client needs. Evaluation includes site visits to Work-life providers and facilities.

Drug-Free Work Place

The EAP offer a needs assessment to determine

- what components of a Drug-Free Workplace are most appropriate for the host or customer organization; and
- for which of the identified components the EAP will be providing services.

Organizational Development/Consultation

It may be said that occupational social work, being an integral part of an organizational conception, represents an important connector between staffing and occupational policy and from its specific perspective may contribute further organizational development and personnel service both to the benefit of the company as well as of the employee's well-being and health preservation.

The offer of assistance has to cover a broad range of services going far beyond simple casework counseling, in order to do justice to the complexity of problem solution strategies, which are appropriate for the respective organization. Occupational social work must be tailored to the organization and its employees according to their specific requirements. Furthermore it may assist and cooperate in the development if personnel and organizational procedures.

Target Groups

Occupational work is intended for

- all employees of the company/authority, as well as their family members,
- superiors and managers,
- the company/authority as an organization and its units.

CLINICAL SERVICES

Assessment

EAP staff should conduct an assessment to identify personal or work problems of employee or covered family members, develop a plan of action and provide, recommend or refer the client to an appropriate resource for problem resolution. The intent is to match clients with the appropriate level of care.

EAPs may be organized in such a way that they act essentially as a point of initial assessment and referral to care givers in the community, be they medical doctors, specialist in alcohol and drug counseling, treatment and rehabilitation, or community-based organizations, including those of a self-help nature. Some EAPs, however, also engage appropriate personnel to provide actual counseling treatment and rehabilitation services for individuals with alcohol- and drug-related problems, provided that referral to outside professionals and agencies are made as necessary.

Short-Term Counseling

EAPA expects standards that will/are to establish procedures to determine if and when to provide short-term (session limited) problem resolution services. Initial assessments shall be conducted by those trained to determine the appropriateness of this kind of invention and match the employee/client with the most suitable resource internally or externally.

Monitoring and Follow-Up

EAPA expects standards that programs will have systems to offer the appropriate monitoring of progress for all clients referred to services external to the EAP.

Referral

EAP staff refer clients to individuals or organizations that offer professional support, advice and treatment in various fields of relevance that match best with the client and his/her needs.

The EAP has procedures to facilitate client referrals, which address the provision of consultation between the EAP and the host or customer organization, and responsibilities for providing follow-up, aftercare, and transition for clients served.

AGREEMENTS

Subcontract/Contractual Agreement

Agreements with subcontractors require the same quality and level of staff training as that of the EAP.

Affiliate Agreements

Affiliate agreements are comprehensive and address the same performance standards required of EAP staff members, such as training and credentials and roles and responsibilities of the EAP and their affiliates.

QUALITY ASSURANCE

Evaluation of Performance

EAPs should evaluate appropriateness, effectiveness and efficiency of their operations. An EAP depends upon having measurable program objectives and quality data control mechanisms. The program evaluation has multiple purposes:

- Documenting the benefits of costs and resources expended on the company.

- Focusing the EAP on employee and organizational needs.
- Continuously improving the quality if the EAP and the efficiency of EAP operations.

Outcome Measurements

In each of its programs, and on an ongoing basis, the EAP measures service outcomes for all clients, including individual client satisfaction with all services receives, level of function level of achievement for goal of assessed problem or request for services.

EAP consumers give a satisfaction questionnaire to each client after the first session in all services.

For its clinical services, the EAP collects and analyzes client self-reported outcomes and counselor reported outcomes toward achieving the service goals established at the first session.

After each episode of training or consultation, the EAP surveys the services recipient to evaluate the impact of the raining or consultation on his/her ability to handle the situation or situations in the workplace that the training or consultation was intended to address.

Return on Investment (ROI)

From the beginning of the program, a method should be developed to measure the cost benefit of the EAP program.

Quality Improvement Affiliates

An EAP that purchases services from affiliates monitors and evaluates those contracted services and implements corrective action, if necessary.

CONCLUSIONS

It will be feasible to achieve accreditation, and it will benefit a program. Using the worldwide guidelines could be another approach, but this entails refining them into a self-study procedure–perhaps with ISO, but that would require funding to accomplish. The US government funded this author's adaptation and development of the first COA standards. Regardless of how accreditation takes place, it is important for the individual EAP program and most critically for the development of the EAP field into a profession.

NOTES

1. These data were presented at the annual Employee Assistance Professional's Association conference in New Orleans, LA (for international section).

2. © 2003, all rights reserved.

The Future of Credentialing
and Accreditation
in Employee Assistance Programs

Louise Hartley
Donald G. Jorgensen

SUMMARY. This article, written by the current presidents of the Employee Assistance Society of North America (EASNA) and the Employee Assistance Professionals Association, examines future issues facing employee assistance programs and discusses the value and relevance of both program accreditation and individual practitioner certification. *[Article copies available for a fee from The Haworth Document Delivery Service: 1-800-HAWORTH. E-mail address: <docdelivery@haworthpress.com> Website: <http://www.HaworthPress.com> © 2003 by The Haworth Press, Inc. All rights reserved.]*

KEYWORDS. Credentialing, accreditation, employee assistance programs, EACC, CEAP, COA, EAPA, EASNA

Louise Hartley, PhD, CPsych, is President, Employee Assistance Society of North America, and Vice President, Family Services Employee Assistance Programs, 2 Carlton Street, Suite 1005, Toronto, ON M5B 1J3, Canada (E-mail: lhartley@fseap. com). Donald G. Jorgensen, PhD, CEAP, is President, Employee Assistance Professionals Association, and President, Jorgensen/Brooks Group, 2120 W Ina Road, Suite 204, Tucson, AZ 85741 (E-mail: don@jorgensenbrooks.com).

[Haworth co-indexing entry note]: "The Future of Credentialing and Accreditation in Employee Assistance Programs." Hartley, Louise and Donald G. Jorgensen. Co-published simultaneously in *Employee Assistance Quarterly* (The Haworth Press, Inc.) Vol. 19, No. 1, 2003, pp. 87-92; and: *Accreditation of Employee Assistance Programs* (ed: R. Paul Maiden) The Haworth Press, Inc., 2003, pp. 87-92. Single or multiple copies of this article are available for a fee from The Haworth Document Delivery Service [1-800-HAWORTH, 9:00 a.m. - 5:00 p.m. (EST). E-mail address: docdelivery@haworthpress.com].

http://www.haworthpress.com/web/EAQ
Digital Object Identifier: 10.1300/J022v19n01_07

INTRODUCTION

A key aspect to establishing and maintaining a credible professional service is how the providers of this service ensure quality. As the employee assistance field has evolved into a full-fledged professional service, it has embraced both credentialing of individual service providers and program accreditation as two complementary ways of assuring the public that they are receiving quality services. The Certified Employee Assistance Professional (CEAP^tm) identifies employee assistance practitioners who meet established standards for competent, workplace and client-centered practices and adhere to an enforceable code of professional and ethical conduct. Program accreditation through an independent organization that establishes and promotes best-practice standards for the employee assistance field gives an organization that meets these stringent standards the ability to market themselves as quality providers.

Currently the employee assistance field has two primary associations that represent the providers of employee assistance services–the Employee Assistance Society of North America (EASNA) and the Employee Assistance Professionals Association (EAPA). As these two organizations discuss the feasibility of creating one new strong clear voice for the industry they are united in their commitment to ensuring quality standards for the individual employee assistance providers and for organizations that deliver employee assistance services. This article will describe the credentialing and accreditation processes that have been supported by each association and future directions for both processes.

PROFESSIONAL CERTIFICATION

Established in 1986, the Employee Assistance Certification Commission (EACC) administers the CEAP. Each candidate must meet experience, professional development and advisement requirements and pass a qualifying examination. The Commission remains the autonomous credentialing body responsible for all aspects of the CEAP program, including establishing policies and procedures of the CEAP credential; examination development; and ethics code enforcement. In January 1987, the EACC commissioners approved the CEAP© designation for those successfully completing certification requirements. As of December 2001, more than 5,500 CEAPs were practicing in the United States and in 16 foreign countries.

FUTURE TRENDS IN EMPLOYEE ASSISTANCE PROFESSIONAL CERTIFICATION

The EACC and the CEAP credential continue to evolve. A new version of the CEAP examination debuted in May 2002 and reflected the revised job requirements identified by a recent role delineation study, and now includes test questions based on different cognitive levels. Application and analysis skills are integral to competency in employee assistance practice, and items to assess these two higher cognitive levels are now incorporated. Already widely recognized in the U.S. and Canada, to make certification more accessible to those in the international community, the Commission has adopted revised advisement requirements and developed internationally relevant versions of the examination leading to the CEAP-I credential.

In response to the employee assistance profession's increasingly sophisticated professional development needs, the EACC continues to prepare for advanced certification (e.g., a Masters-level CEAP) and the possibility of providing subspecialty credentials, such as a substance abuse professional and critical incident stress debriefing certification. The U.S. Army recently adopted the CEAP credential for its civilian and uniformed employees working as substance abuse specialists.

EMPLOYEE ASSISTANCE PROGRAM ACCREDITATION

Accountability for quality of employee assistance service delivery has driven the need for employee assistance program accreditation. In 1981, the Standards for Employee Alcoholism and/or Employee Assistance Programs were drafted by a joint committee representing the Association of Labor/Management Administrators and Consultants on Alcoholism (ALMACA–and predecessor to EAPA), the National Council on Alcoholism, the Occupational Program Consultants, the National Institute on Alcohol Abuse and Alcoholism. From this early effort until today, four groups have remained involved in program standards.

1. The Employee Assistance Professional Association (EAPA) originally published its Professional Standards for employee assistance programs in 1988, based on the widely accepted employee assistance Core Technology. Part Two, employee assistance Professional Guidelines, was published in 1992. Parts Three and Four, Glossary of employee assistance Terms and the employee assistance Program Self-Evaluation, were published in 1994. Many employee assistance programs have based their operating procedures and best practices on these standards.
2. The Council on Accreditation of Rehabilitation Facilities (CARF) established an employee assistance program accreditation product in 1988. Most of the 21 EAPs receiving CARF accreditation to date are provided as part of substance abuse treatment programs or community human service organizations.
3. The Council on Accreditation for Children and Family Services (COA) began accrediting Employee Assistance service providers as components of Child and Family Service agencies in 1987. Of the 100+ employee assistance program services accredited under this process, most are part of larger multiservice child and family agencies.
4. The Employee Assistance Society of North America (EASNA), founded in 1985, established a peer-reviewed accreditation program in 1990. Thirty-five programs were accredited under this system, most of them from Canada. Subsequently, Canadian employee assistance requests for proposals have routinely inquired about EASNA program accreditation.

As consolidation between U.S. and Canadian EAPs began in the 1990s, interest by U.S. firms in program accreditation has heightened.

EASNA and COA subsequently formed a partnership for development of a joint accreditation product in 2000, the EASNA EAP Accreditation Program as administered by COA©. Developmental support for the review of best practices in developing the accreditation came from the U.S. Department of Health and Human Services' Substance Abuse and Mental Health Services Administration. This latest iteration of employee assistance program accreditation has consolidated the existing EASNA and COA processes and includes new standards regarding emerging employee assistance practice areas such as work-life programs, organizational development, drug-free workplace services, legal services, Internet and telephonic services, and international services. After beta testing at five unique employee assistance program sites, the revised accreditation procedures were published in July 2001 and now are solely administered by COA.

EAPs may undertake a program accreditation for a variety of reasons. Accreditation is an assurance to the community and an organization's clients that it has met recognized standards of operation and professional conduct. An internal employee assistance program may seek accreditation to benchmark their performance, and the results might serve as a hedge against outsourcing. External providers may seek accreditation to gain a marketing edge. Employee assistance training departments may use accreditation findings as a justification for increasing resources or raising supervisory levels to meet accreditation criteria.

Thus, voluntary program accreditation programs provide external, objective benchmarks for consumers and stakeholders of recognized standards of practice and quality of service provision. Additionally, accreditation findings provide employee assistance staff a detailed analysis of specific strengths and weaknesses in the organization's governance, operations and provide a blueprint for ongoing improvement (EASNA, 1990).

CONCLUSION

The employee assistance industry is at another crossroads in its development. Employee engagement and retention are key workplace issues in which the employee assistance industry can play a critical role. A cornerstone to presenting a credible business case for employee assistance service providers to be part of the solution to these current workplace issues is the quality foundation that is built through credentialing and accreditation. Thus, as EASNA and EAPA dialogue about how to create one united strong association for the industry, they both agree that the quest for quality assurance is best provided by processes that both credential professionals delivering employee assistance services and accredit quality employee assistance programs.

APPENDIX

Council on Accreditation Employee Assistance Program Accreditation Standards (2nd Edition)

This Appendix provides an excerpt from the Council on Accreditation's *Employee Assistance Programs Standards and Self-Study Manual, 2nd Edition.* For a full text version of the Manual, including Evidence of Compliance for the standards, please contact COA at 866.262.8088.

I. ADMINISTRATION AND MANAGEMENT

I.1   Legal Compliance

I.1.01   The EAP complies with all applicable federal, state or provincial, and local laws and regulations.

[Haworth co-indexing entry note]: "Council on Accreditation Employee Assistance Program Accreditation Standards (2nd Edition)." Co-published simultaneously in *Employee Assistance Quarterly* (The Haworth Press, Inc.) Vol. 19, No. 1, 2003, pp. 93-190; and: *Accreditation of Employee Assistance Programs* (ed: R. Paul Maiden) The Haworth Press, Inc., 2003, pp. 93-190.

I.1.02   The EAP does not possess any outstanding work orders, notices of violation, or negative directives from any governmental or quasi-regulatory body.

I.1.03   The EAP complies with all self-reporting requirements associated with licensure, accreditation, and/or other appropriate external review bodies.

I.2   *Legal Structure*

I.2.01   The external EAP or the internal EAP's host organization is legally authorized to operate as one of the following:

a. a for-profit organization that is organized as a corporation, partnership, sole proprietorship, or association, and has a duly promulgated charter, articles of incorporation, partnership agreement, articles of association, constitution, and/or bylaws;

b. a not-for-profit organization that is organized as an identified sub-unit of another legal entity recognized under state or provincial law; or

c. a not-for-profit organization that is incorporated in the state or province in which it operates or is headquartered, and has a duly promulgated charter, constitution, and/or bylaws with its own board of directors.

I.2.02   The EAP's charter or constitution provides that disposition of assets upon dissolution of the corporation shall be in keeping with the purpose of the EAP and comply with applicable legal and contractual requirements.

I.3   *Organization of the Board of Directors*

Interpretation (I.3):
  Please note that a board of directors/trustees acts as the governing body or "owners" of a not-for-profit EAP. In a for-profit EAP, the gov-

erning body or "owners" are the EAP's shareholders. Please apply the appropriate perspective when completing these standards.

I.3.01   The EAP maintains one or more bodies, such as a board of directors/trustees, or advisory board, that regularly advise the EAP on its policies, management, planning, finances, use of resources, and service delivery.

Interpretation (I.3.01):

Most privately-held for-profit organizations demonstrate compliance with this standard by establishing an advisory board. Not-for-profit or publicly traded for-profit organizations will often demonstrate compliance through a board of directors/trustees.

I.3.02   The board of directors is sufficient in size and structure to:

    a. engage in long-term planning;
    b. develop and adopt policy;
    c. develop resources; and
    d. provide financial oversight.

Interpretation (I.3.02):

A large organization is likely to have an elaborate committee or task force structure to accomplish these goals, whereas a small organization may not require such a framework. As long as the EAP can demonstrate that the board of directors carries out its responsibilities effectively and thoroughly through regular meetings and clear responsibilities, compliance will be achieved.

I.3.03   The bylaws, constitution, or similar legal document of the publicly traded or not-for-profit EAP is reviewed every four years and establishes:

    a. the structure, size, and responsibilities of the board of directors;

b. the minimum number of board of directors' meetings and quorum requirements;

c. the body to which the board of directors delegates interim authority; and

d. a process for assessing and implementing board responsibilities, such as establishing task forces/committees, and respective responsibilities and composition.

I.3.04   The documents listed in I.3.03 also set forth the:

a. eligibility requirements for board of directors membership;

b. mechanisms for recruitment, selection, rotation, and duration of board of directors membership; and

c. mechanisms for election of officers and the duration of officer terms.

I.3.05   All members of the EAP's board of directors:

a. receive an orientation to the responsibilities of membership;

b. receive a manual with current, relevant organizational material that specifies their fiduciary and other responsibilities to the EAP;

c. receive a formal orientation to the EAP's mission, history, goals, objectives, structure, methods of operation, and introductions to key staff; and

d. are familiarized with the activities of the organization through a visit to the EAP.

Interpretation (I.3.05):

"Relevant organizational materials" noted in (b) include, but are not limited to, bylaws, mission statement, and relevant policies and procedures.

I.4 *Owners/Senior Management or Board of Directors' Responsibilities*

I.4.01 The owners/senior management or board of directors assumes responsibility for setting the EAP's long-term direction.

I.4.02 In fulfilling its oversight responsibilities, the owners/senior management or board of directors:

a. ensures that all planned or provided services are consistent with the EAP's mission and long-term plan; and
b. determines whether services are within the EAP's capabilities and resources.

Interpretation (I.4.02):
The owners/senior management or board of directors' effectiveness in these oversight responsibilities is directly linked to the EAP's quality improvement processes and long-term planning described in VII. Quality Improvement.

I.4.03 The board of directors assumes responsibility for policy development and maintenance by:

a. adopting policies;
b. reviewing policies at specified intervals and whenever legal requirements or regulations change; and
c. approving any changes to policies resulting from recommendations or negotiation with a recognized collective bargaining unit.

Interpretation (I.4.03):
An organization that follows a policy governance model may not typically develop, ratify, and maintain statements known as "policies." However, statements that are distillations of organizational principles, philosophies, practice, or "ends" may be considered policies for the purposes of this standard.

The standard requires that the board of directors actively exercises its policy-setting prerogative, i.e., policies are periodically reviewed as a whole, and specific policy matters regularly receive the board of directors' attention. The board of directors must view policy as the board of directors' major means of providing a framework and guidance for the EAP's overall direction.

I.4.04   The EAP maintains a comprehensive policies and procedures manual that includes owners- or board of directors-approved policy statements.

Interpretation (I.4.04):

In an internal EAP, the policies and procedures manual may be part of the organization-wide policies and procedures manual.

I.4.05   The EAP's owners/senior management or board of directors' fiscal responsibilities include:

a. reviewing and approving the EAP's annual budget;
b. reviewing fiscal summaries at least quarterly to examine the relationship of the budget to actual expenditures and revenues;
c. examining fiscal policy and the recommendations of the EAP's auditors; and
d. annually evaluating the chief executive officer or equivalent's management of the EAP's fiscal affairs.

Interpretation (I.4.05):

In an internal EAP, the activities described in this standard may occur as part of organization-wide management and/or board functioning.

I.4.06   The board of directors and/or advisory board maintains up-to-date minutes and records generated from all meetings.

Interpretation (I.4.06):

In EAPs with both a board of directors and an advisory board, gaps or lapses in advisory board records will not jeopardize the EAP's compliance.

I.5 Owners/Senior Management or Board of Directors' Risk Management Responsibilities

I.5.01 The EAP reports to the board of directors or its designated authority on the nature of risks and actions taken to address them.

I.5.02 The owners/senior management or board of directors reviews patterns of complaints and grievances and addresses specific problematic or unresolved issues that may expose the EAP to liability.

I.5.03 The owners/senior management or board of directors ensures that the EAP complies with all laws related to fiscal accountability and governance.

I.6 Board of Directors' Responsibilities Related to the Chief Executive Officer

I.6.01 The board of directors:

a. appoints a chief executive officer and delegates authority and responsibility for the EAP's management and implementation of policy; and
b. holds the chief executive officer accountable for the EAP's performance.

I.6.02 The EAP's board of directors:

a. evaluates the chief executive officer's performance in writing at least annually against established performance criteria that are linked to the EAP's long-term plan; and

      b. ensures that the chief executive officer participates in the evaluation process and reviews, signs, and responds to the evaluation before it is entered into his/her record.

I.6.03   The EAP's board of directors reviews the fairness of the chief executive officer's compensation and benefits on an annual basis in relationship to industry practices and federal law.

I.6.04   In the absence of the chief executive officer, the EAP's board of directors:

      a. has a written plan for delegating authority;
      b. designates an interim chief executive officer, if necessary;
      c. charges a committee with responsibility for conducting a formal search, where necessary; and
      d. provides the resources needed to carry out the search effectively.

I.7   *Chief Executive Officer*

Interpretation (I.7):

In a large organization, the chief executive officer's functions, as described in these standards, may be assumed in part by a designee of the chief executive officer, such as an administrator, director, or president.

I.7.01   The chief executive officer or his/her designee:

      a. plans and coordinates the development of policies governing the EAP's services with the owners or board of directors; and
      b. attends all meetings of the board of directors and advisory board, with the possible exception of those held for the purpose of reviewing the executive's performance, status, or compensation.

Interpretation (I.7.01):

The chief executive officer or his/her designee involves, consults, and gives leadership to the board of directors in the planning, policy, and decision-making processes. The chief executive officer or his/her designee and board of directors work as an effective team: information, coordination, staffing, and assistance are provided by the executive to support the board of directors in its policy making and oversight functions.

I.7.02   The chief executive officer or his/her designee provides written, comprehensive reports to the board of directors according to a mutually agreed upon schedule regarding:

     a. present financial status and anticipated problems;
     b. financial planning and funding alternatives;
     c. the operation of present programs, including areas of non-compliance with EAP policy;
     d. the implementation and annual review of long-term plans; and
     e. any other issues related to the EAP's achievement of its mission.

Interpretation (I.7.02):

Such reports must be provided at least annually. Reports must be responsive to the board of directors' need for information upon which to base decisions about short-term financial and budgetary matters, and to plan for the near term.

I.7.03   The chief executive officer or his/her designee's primary responsibility is management of the EAP and s/he:

     a. obtains board of directors' approval for employment activities outside of the EAP; and
     b. assumes no duties that are unrelated to and/or interfere with his/her management responsibilities.

I.7.04   The chief executive officer or his/her designee reviews policies and procedures annually for their continued applicability.

I.7.05   The chief executive officer or his/her designee ensures that human resources management complies with federal and state or provincial employment law.

I.8   *EAP Policies*

I.8.01   The EAP defines itself in its policy and promotional materials as either an internal, external, or internal/external EAP.

I.8.02   The internal EAP operates as a distinct service within the host organization.

I.8.03   EAP policy establishes that its service delivery is tailored to the needs and requests of its customer organizations as set forth in its contracts with the customers.

I.8.04   EAP policy sets forth its relationship with any managed behavioral healthcare companies that provide services to the host or customer organization.

I.8.05   EAP policy establishes that it will not withdraw services prior to the number of sessions stated in its contract with the client, unless the client:

a. requires longer term service and needs an immediate referral;
b. does not wish or need to continue service; or
c. is unable to continue service.

I.8.06   The EAP's policies, procedures, and practices prohibit discrimination in service delivery.

Interpretation (I.8.06):
In an internal EAP, the nondiscrimination policy may be part of the organization-wide policies and procedures manual.

I.9   *EAP Service Design*

I.9.01   The EAP has a program description that includes:

    a. an explanation of the kinds of services offered by the EAP, including the availability of emergency services;
    b. the kinds of referrals the EAP accepts and the referral process to the EAP;
    c. the nature and extent of preventive and health promotion activities to be undertaken through the EAP;
    d. the location of the EAP services; and
    e. ways to access EAP services.

I.9.02   The program description also includes:

    a. a clear description of any mandated disclosures;
    b. a statement verifying that the EAP adheres to all legal requirements; and
    c. a statement maintaining the EAP's neutrality with the host or customer organization with respect to employee/employer relations.

I.9.03   The EAP service is designed to:

    a. help organizations develop and maintain optimum work environments for their employees;
    b. help employees become more productive at work;
    c. help employees and eligible participants with relationship, family, addiction, legal, emotional, stress, work-life balance, and other personal problems;

d. make referrals for individuals who have severe psychological problems and substance abuse disorders that require outpatient, partial, or residential treatment; and

e. provide preventive strategies aimed at stimulating employee awareness and education to encourage early intervention.

Note: As used in this standard, "addiction" addresses alcohol, drugs, gambling, sexual, internet, and other addictions or addictive behaviors.

I.9.04   The EAP provides the following services:

a. assessment and referral;

b. employee education and outreach;

c. training for supervisors, managers, and human resource and union representatives;

d. management/supervisory consultation;

e. an EAP referral network for obtaining needed services that are not provided under the contract and/or are not available at the EAP; and

f. follow-up on referrals.

II. MANAGEMENT OF EAP HUMAN RESOURCES

II.1   Human Resources Planning, Organization, and Deployment

II.1.01   The EAP:

a. assesses the type and number of personnel required to accomplish its mission, goals, and objectives;

b. establishes goals for retention of personnel;

    c. measures and evaluates the rate of personnel turnover against benchmarks established for each job category; and

    d. takes prompt action to correct identified job retention and satisfaction problems.

II.1.02   In determining and reviewing the size of staff member work-loads, the EAP assesses:

    a. the work and time required to accomplish assigned tasks and job responsibilities;

    b. service volume, accounting for assessed level of needs of new and current clients and referrals; and

    c. standards of best practice, where they exist.

II.1.03   The EAP retains sufficient numbers of qualified individuals to:

    a. efficiently and effectively meet the demand for all services it provides; and

    b. provide and coordinate the services that are within the EAP's scope and resources.

Interpretation (II.1.03):

    These individuals include those available on a full-time, part-time, or contractual basis.

II.2   Human Resources Practices

II.2.01   The EAP maintains an organizational chart that specifies all staff positions and, if applicable, the position of the EAP within the host organization.

II.2.02   Personnel policies and procedures are outlined in a manual, handbook, or other document that is provided to all EAP staff and which covers, but is not limited to:

    a. personnel practices;

    b. working conditions (e.g., hours of operation, fire protocol);

    c. training and development opportunities;

    d. conditions and procedures regarding disciplinary actions;

    e. insurance protections; and

    f. expectations for staff conduct and performance.

II.2.03   The human resources procedures manual also addresses the following:

    a. benefits;

    b. paid and unpaid leave, holidays, and other time off;

    c. wage and promotion policies; and

    d. work-life enhancement programs, if applicable.

II.2.04   The EAP requires personnel to sign or initial a statement indicating that they have received and understand the policies and procedures contained in the manual.

II.2.05   The EAP maintains an individual personnel record for each staff member that includes:

    a. a job description;

    b. the staff member contract or letter of hire;

    c. a resume;

    d. a copy of current licenses;

    e. malpractice insurance coverage;

    f. a signed confidentiality statement; and

    g. other appropriate documents.

II.2.06   Staff have access to their own records and may review, add, and correct information contained in them, in accordance with applicable state or provincial law.

II.2.07   EAP procedures address:

a. the use of off-site staff members; and
b. appropriate work environments used in relation to their job responsibilities.

II.3 Human Resources Policies

II.3.01 The EAP's personnel policies and procedures state that it will not discriminate against any person or categories of persons protected by applicable federal, state or provincial, and/or local laws.

Interpretation (II.3.01):

When recruitment and hiring criteria include consideration of specific protected characteristics, such as gender, religion, and national origin, the EAP should seek legal advice as to whether these characteristics are "bona fide occupational qualifications" that are relevant to the EAP's normal operation. Specifically, such organizations that operate in the United States should seek legal advice regarding the applicability of sections 702 and 703 of Title VII of the Civil Rights Act.

II.3.02 The EAP complies with applicable laws and regulations governing fair employment practices and contractual relationships.

Interpretation (II.3.02):

This standard applies to the EAP's employment practices and contracts, including its use of independent contractors and temporary staff members, and its compliance with collective bargaining agreements. The major federal laws in the United States that govern employer/employee relations include, but are not limited to: the Civil Rights Act of 1964 (as amended by the Equal Employment Opportunity Act of 1972 and the Civil Rights Act of 1991), the Fair Labor Standards Act, the Equal Pay Act, the Age Discrimination in Employment Act, the Americans with Disabilities Act, the Family and Medical Leave Act, the Occu-

pational Safety and Health Act, the National Labor Relations Act, as well as the regulations implementing all of the above statutes, and Executive Order 11246. The major federal laws in Canada that govern employer/employee relations include, but are not limited to: all chapters of the Canada Labour Code, the Fair Wages and Hours of Labour Act, the Canadian Human Rights Act, the Employment Equity Act, the Employment Insurance Act, and implementing regulations for all of these statutes. Fair employment practices will be addressed for the laws of each EAP's country of operation.

II.3.03   The EAP's equal employment opportunity or employment equity policy:

    a. clearly states its practices for recruitment, employment, transfer, and promotion of personnel;
    b. is appropriately disseminated to staff; and
    c. is used in recruitment processes.

II.3.04   All staff members receive a copy of the EAP's grievance procedures which include:

    a. the steps for personnel to lodge complaints, grievances, and appeals;
    b. requiring a timely written response to complaints, grievances, and appeals according to its stated procedures;
    c. documenting responses and actions taken in a manner consistent with stated procedures;
    d. informing the aggrieved staff member of the complaint's resolution; and
    e. maintaining a copy of the notification of resolution in the personnel record.

Interpretation (II.3.04):
  Implementation of the standard requires appropriate training for staff on the EAP's grievance procedures.

II.3.05   The EAP's personnel policies prohibit nepotism and specify:

    a. conditions for employing and retaining relatives of board of directors members;
    b. conditions for employing and retaining relatives of staff members;
    c. conditions for using relatives and business partners for referrals; and
    d. protection against favoritism in supervision and promotion.

Interpretation (II.3.05):

It is permissible for personnel or relatives of personnel to be members of the board of directors or advisory board, provided that such representation does not undermine the board of directors or advisory board's independence and diversity, and provided that relatives do not work within the same hierarchy of supervision. "Relatives" include those persons related to staff members, the board of directors or the owners, through family of origin, extended family, or marital affiliation.

II.3.06   The EAP's policies on harassment include:

    a. a clear definition of the kinds of behavior the EAP recognizes as harassment;
    b. a statement that the EAP will have zero tolerance for prohibited harassment;
    c. a prohibition against personnel harassing clients, supervisees, colleagues, or other persons or groups with whom they have contact as representatives of the EAP;
    d. the EAP's commitment to take necessary and appropriate action to prevent or eliminate harassment on the job; and
    e. procedures for reporting harassment to management.

Interpretation (II.3.06):

The EAP's harassment procedures should allow personnel to bypass any person in the reporting process who is also the alleged harasser.

Implementation of the standard requires appropriate training for staff on the EAP's harassment policies.

II.3.07   The EAP that provides services in the United States establishes a policy that addresses its role and responsibilities in relation to providing services to employees of host or customer organizations that require drug testing.

II.4   Recruitment and Selection Practices

II.4.01   The EAP maintains a detailed job description or description of responsibilities for all staff members.

II.4.02   The EAP employs only those persons who are qualified according to the job description and selection criteria for the positions they occupy.

Interpretation (II.4.02):
All personnel have at least the requisite stated qualifications in their written job description, and professional personnel have the appropriate entry-level degree for their respective profession or discipline.

Note: This standard applies only to personnel.

II.4.03   Screening procedures for new staff members require, unless contravened by law:

   a. review of state or provincial criminal records;
   b. review of state or provincial child abuse and neglect registries for staff members that serve minors and vulnerable adults; and
   c. review of sex offender registries.

Interpretation (II.4.03):
This standard applies to all programs and other staff, including security officers, who may have contact with clients. The EAP should ensure

*that its practices comply with federal and state or provincial laws re-
garding background checks. For example, EAPs that operate in the
United States must comply with the federal Fair Credit Reporting Act.
If the state or province in which the EAP operates prohibits criminal records checks or civil child abuse and neglect registry checks, the EAP
complies with this standard by providing a copy of such law or regulation.*

II.4.04   The EAP verifies the credentials of all staff members and
maintains documentation on:

a. educational credentials;
b. training;
c. proof of current liability insurance;
d. relevant experience;
e. competence in the required role;
f. recommendations of peers and former employers; and
g. state or provincial registration, licensing, or certification
requirements for their respective disciplines, if any.

Interpretation (II.4.04):
*Information from a national credential verification data bank, re-
spective to the country at issue, is acceptable evidence of compliance. If
a delay in procuring authentication of credentials occurs, such information must be finalized within 90 days of hire, unless such delays are
beyond the control of the EAP.*

II.4.05   EAP policy prohibits permitting personnel or contractors
who have a documented history of assaultive behavior to
have interaction with, or provide oversight to, vulnerable
populations.

Interpretation (II.4.05):
*"Assaultive behavior" includes any offensive touching or threat of
offensive interaction with a vulnerable population, such as children,*

youth, older adults, or impaired adults. Proscribed personnel or contractors include individuals who have been determined by judicial or administrative proceedings to have threatened or harmed a member of a vulnerable population. The standard prohibits the EAP from allowing such individuals to work directly with these populations or within a facility where interaction may occur.

II.5   Human Resources Assessment and Evaluation

II.5.01   The EAP annually seeks input, determines the level of personnel satisfaction, and institutes corrective action regarding the following:

    a. leadership and management;
    b. personnel development, recognition, and career opportunities;
    c. quality of work environment;
    d. adequacy of compensation and benefits;
    e. interdepartmental communication; and
    f. EAP policies and procedures as they are revised.

Interpretation (II.5.01):
Formal procedures include personnel and board of directors representation on a joint committee, separate committees dealing with personnel matters, a task force assigned review responsibility by management, a bargaining committee or union representative in the case of a unionized setting, and/or another functional means of providing the opportunity for personnel to have input on matters which affect them. These efforts are part of the EAP's short-term planning.

II.5.02   The EAP sets salary and benefits scales according to:

    a. prevailing labor market trends for all staff members in comparable settings, in the case of non-unionized settings; and/or

b. the current collective bargaining agreement, in the case of a unionized organization.

II.5.03 The EAP reviews its job descriptions and selection criteria at least every two years to ensure that:

a. education and experience requirements are relevant and appropriate to the EAP's programs, client needs, and specific services provided; and
b. qualifications or credentials are reasonably related to the level of competence required for the tasks involved.

II.5.04 An EAP that uses independent contractors:

a. carefully reviews its retention of independent contractors, such as affiliates and temporary staff members, against the Internal Revenue Service, Canada Customs and Revenue Agency, or other appropriate government or international requirements;
b. conforms with those requirements; and
c. does not place the organization at financial risk by failing to comply with applicable legal requirements.

II.6 *Accountability and Performance Review*

II.6.01 In conjunction with personnel, the EAP develops outcomes-oriented performance expectations for each position, which are discussed with each staff member.

II.6.02 Performance reviews are conducted at least annually between each staff member and his/her supervisor and include:

a. an assessment of job performance in relation to the expectations defined in the job description and the objectives established in the most recent evaluation;

b. clearly defined objectives for future performance; and

c. recommendations for further training and skill building, if applicable.

Interpretation (II.6.02):
Performance reviews are conducted in person.

II.6.03   Staff members are given the opportunity to sign the written performance review, obtain a copy, and provide written comments before the report is entered into their personnel record.

II.6.04   The EAP conducts an exit interview with all personnel who voluntarily leave the organization.

Interpretation (II.6.04):
This interview enables personnel to address administrative issues related to the transition, as well as to provide feedback on the EAP's strengths and weaknesses.

II.7   Affiliate Engagement

II.7.01   The EAP screens all affiliates and verifies credentials in accordance with the requirements of standards II.4.03 and II.4.04.

II.7.02   EAP contract files contain the following documents used in the affiliate selection process:

a. affiliate agreements;
b. job descriptions;
c. resumes;
d. copies of current licenses;
e. malpractice insurance coverage; and
f. signed confidentiality statements.

II.7.03   An EAP that uses the services of professionals on a per-interview, hourly, or independent contractor basis has regular mechanisms to ensure the quality of services provided.

Interpretation (II.7.03):
"Regular mechanisms" may include quality improvement reviews, evaluation of services, or other assessments performed by a third-party.

III. HEALTH AND SAFETY

III.1   Environmental Quality

III.1.01   The physical environment reflects the EAP's commitment to provide comfort and dignity to clients and personnel of diverse backgrounds and ages.

III.1.02   The EAP maintains a service environment in all offices that is safe, clean, free of fire hazards, smoke free, and professional.

III.1.03   The EAP has adequate office space to ensure client and counselor confidentiality.

Interpretation (III.1.03):
Counselor confidentiality may be achieved through a number of different mechanisms, e.g., soundproof rooms or white noise devices.

III.1.04   The organization regularly seeks the input of clients and personnel about the quality of the environment and focuses its efforts on remediating identified problems.

III.2   Accessibility

III.2.01   In planning the location of its offices and branches, the EAP considers:

    a. accessibility, availability, and affordability of public transportation;

    b. location of other community resources; and

    c. special needs of actual and potential clients within the EAP's geographic service areas.

Interpretation (III.2.01):

This standard requires the EAP to address the needs of persons with disabilities.

III.2.02   An EAP office that serves populations with special needs, such as older adults, persons with disabilities, or young children:

    a. designs and adapts its facilities to address the visual, auditory, linguistic, and motor limitations of its service population; and

    b. provides assistive technology, as needed.

Interpretation (III.2.02):

The standard requires the EAP to adapt its environment to the special needs of clients. Accessibility of services is an integral component to meeting need equitably. The EAP should attempt to deploy and adapt its facilities so that they are usable by all those in need and otherwise eligible. This includes providing or arranging for communication assistance for persons with special needs, who have difficulties making their service needs known, by providing assistance such as a computer, telephone amplification, sign language services, or other communication methods to facilitate service.

For EAPs that use affiliates, the EAP shall include the requirements of III.2.02 in the affiliate agreement as stated in V.5.02, and verify compliance through review of VII.6.02.

III.3   *Functional Safety and Compliance with Health and Safety Codes*

III.3.01   The EAP complies with federal and state or provincial, and local legal requirements governing public accessibility, health, and safety.

Interpretation (III.3.01):

Applicable federal laws governing organizations that operate in the United States include Title III of the Americans with Disabilities Act ("ADA"), as well as various state and municipal laws. Applicable laws governing EAPs that operate in Canada include the Charter of Rights and Freedoms, Canadian Human Rights Act, provincial human rights statutes, and provincial and municipal building codes.

III.3.02　For facilities, offices, and grounds that are regularly used, rented, or owned, the EAP maintains a permanent file of reports, including incident reports, that demonstrate its compliance with all:

    a. certification of occupancy requirements;
    b. zoning and building codes;
    c. occupational safety and health administration codes;
    d. health, sanitation, and fire codes; and
    e. all other applicable safety codes.

Interpretation (III.3.02):

Compliance with the standard can be demonstrated through documentation from public or private authorities. For example, some jurisdictions do not make inspection reports available to those who rent rather than own property. In such a case, the EAP may solicit a recognized expert to verify compliance with the law. This interpretation is intended to provide some flexibility for EAPs renting facilities from landlords who will not give them access to needed information. The organization should document its attempts to gain access to the information.

III.3.03　The EAP conducts and documents:

      a. regular inspections and preventative maintenance to ensure the safety of its premises, equipment, and fixtures; and

    b. a monthly review of the physical plant's safety systems including fire safety and fire extinguishers, emergency exits, lighting, and other mechanisms that identify hazardous conditions.

Interpretation (III.3.03):

"Hazardous conditions" considered by the standard include: uncovered electrical outlets; unsecured floor coverings or equipment; walk-in freezers or refrigerators that do not open from the inside; stairs without handrails; harmful water temperatures; inadequacy of light, ventilation and temperature; unscreened areas or unmarked glass doors; and unsafe use of electrical appliances and objects, such as hair dryers, space heaters, radios, or toys that are used by children or others who may be vulnerable.

III.3.04   All EAP offices meet legal requirements for fire drills and inspections.

III.3.05   The EAP follows anti-crime procedures to ensure that all buildings, grounds, and facilities are safe and secure for clients and personnel.

Interpretation (III.3.05):

The EAP must tailor security measures to address high-risk environments, e.g., those posed by late evening hours or face-to-face counseling with volatile clients. Possible ways to address the standard include the use of dead-bolt doors, panic alarms, entrance bells, congregate working areas to improve safety, and other security linkages. Such procedures may include measures against vandalism and bars on windows.

III.4   *Emergency Response*

III.4.01   Procedures for responding to accidents, fire, medical emergencies, water emergencies, natural disasters, and other life threatening situations:

    a. address the needs of persons with special needs;

    b. specify evacuation procedures and appropriate responses to medical emergencies;

    c. address voluntary or involuntary closure of facilities in emergency situations; and

    d. require notifying the client's parent or legal guardian, if appropriate, and other appropriate authorities.

III.4.02   The EAP establishes policies that address workplace violence, including responding to emergency situations within the workplace, and the training of staff in these areas.

Interpretation (III.4.02):

Workplace violence includes situations that involve a threat or actual harm or violence to personnel or clients, including hostage situations, bomb threats, unlawful intrusion, or assaultive behavior.

III.4.03   The EAP has a plan for handling workplace emergencies that includes the establishment of:

    a. a temporary work site;

    b. a computer data recovery plan;

    c. emergency telephone, internet, and facsimile use;

    d. procedures for handling the media; and

    e. a system for communicating with the board of directors, personnel, clients, and the public.

IV. FINANCE

IV.1   Financial Planning

IV.1.01   The EAP has a budget that serves as a plan for managing its financial resources for the fiscal year and includes provisions for:

    a. staff training;
    b. performing evaluation activities;
    c. administrative overhead;
    d. travel, if appropriate; and
    e. staff salaries.

IV.1.02   The owners/senior management or board of directors:

    a. reviews and approves all planned deviations of signifi-cance from, and revisions to, the budget prior to imple-mentation; and
    b. ensures that budget-to-actual variance analyses are per-formed after year-end numbers are finalized.

IV.1.03   At least monthly, the EAP conducts an analysis of financial performance against budgeted projections.

IV.1.04   Annually, the EAP conducts an inventory of significant assets, including securities, and compares them with permanent records.

IV.2   *Financial Information*

IV.2.01   External EAPs have sufficient financial resources to cover operational expenses for the next 12 months.

Interpretation (IV.2.01):
    "Financial resources" include cash and/or a credit line.

IV.2.02   As part of its short-term planning process, the EAP analyzes in-formation on service revenues and actual service delivery costs per client served on a per capita basis to evaluate effec-tiveness and efficiency of services.

Interpretation (IV.2.02):
    The analysis can be done independently or with an outside entity.

IV.2.03   The EAP bills its customers on a per capita basis and provides its customer organizations with a copy of its procedures.

Interpretation (IV.2.03):

The cost structure and pricing of EAP services for customers is based on the following:

- *projected utilization rates;*
- *number of counseling sessions provided;*
- *additional training/consulting services provided;*
- *educational level of counselors;*
- *special staff requirements, e.g., bilingual abilities;*
- *financial impact of traumatic/change events on the EAP;*
- *cost per hour for counseling or cost per client, if appropriate;*
- *audit/evaluation requirements, if appropriate;*
- *unusual travel requirements;*
- *costs per hour for account management;*
- *costs related to MIS, and other overhead expenses;*
- *costs related to providing telephone and online services, if appropriate; and*
- *costs related to subcontracts.*

IV.3   Fiscal Management System

IV.3.01   Annual financial statements are prepared in accordance with Generally Accepted Accounting Principles.

IV.3.02   The EAP maintains an accounting ledger that addresses the timely payment of accounts payable and collection of accounts receivable.

IV.3.03   The EAP acts in accordance with an internal accounting control system that addresses:

a. prevention of error, mismanagement, or fraud;
b. an inclusive and descriptive chart of accounts;
c. prompt and accurate recording of revenues and expenses;
d. control over prompt payment of expenditures; and
e. information on all funds, including each fund's source and pertinent regulations governing each fund.

Interpretation (IV.3.03):
Compliance with this standard can be demonstrated through mandatory periods in which accounts are controlled by a staff member other than the primary accountant, or by having two individuals responsible for the accounts.

IV.3.04   The EAP's accounting procedures also address:

a. safeguarding and verifying assets;
b. separation of duties to the extent possible; and
c. disbursement and receipt of monies.

Interpretation (IV.3.04):
Please note that these procedures must address cash, checks, and other accounts.

IV.3.05   The EAP seeks to conserve its fiscal resources by:

a. taking advantage of benefits allowed tax-exempt organizations, where applicable;
b. maintaining sound policies regarding purchasing and inventory control; and
c. using competitive bidding, when applicable, according to board of directors policy and law or regulation.

IV.3.06   Accounting records are kept up-to-date and balanced on a monthly basis, as demonstrated by:

a. reconciliation of the bank statement to the general ledger;
b. reconciliation of subsidiary records to the general ledger;
c. up-to-date posting of cash receipts and disbursements;
d. monthly updating of the general ledger; and
e. review of the bank reconciliation by at least two personnel, one of whom is not involved in maintaining the accounting records.

Interpretation (IV.3.06):
Subsidiary records include, but are not limited to: Accounts Receivable, Accounts Payable, and Property, Plant, and Equipment.

IV.3.07   Oversight and management of the EAP's accounting system requires that:

a. a fiscal officer or business manager who is responsible for maintaining the financial accounts has prior accounting experience, and an accounting degree or CPA credential, as appropriate to the EAP's size and complexity;
b. all personnel who use the accounting system are oriented to the system and are retrained regarding any changes; and
c. internal control systems are managed or reviewed by more than one person.

IV.3.08   As required by law, the EAP makes timely payments to, or provides proof of exemption from, the appropriate taxing authorities.

Interpretation (IV.3.08):
For EAPs operating in the United States, taxing authorities include the Internal Revenue Service, state and local tax bodies, FICA, and property tax assessors. For EAPs operating in Canada the taxing authority is the Canada Customs and Revenue Agency.

IV.4   Financial Accountability

IV.4.01   The EAP makes summary information regarding its financial status available for inspection by external reviewers.

IV.4.02   An EAP that annually reports its revenues at or in excess of $500,000, or is otherwise required, undergoes an audit of its financial statements and such an audit is performed:

a. within 180 days of the end of its fiscal year; and
b. by an independent, certified public accountant who is approved by the owners or board of directors.

Interpretation (IV.4.02):
Organizations in the United States receiving in excess of $300,000 in federal funds must perform an audit to comply with the requirements of the Single Audit Act, 31 U.S.C. §§ 7501 et. seq. Note that many organizations are required to perform an audit to receive grant monies, lines of credit, or other third-party funding.

IV.4.03   In EAPs that conduct an audit, the owners or board of directors:

a. review the findings, the accompanying financial statements, and any management letter that may accompany the audit report;
b. accepts the audit; and
c. ensures that management promptly acts on the recommendations of a management letter, if any.

Interpretation (IV.4.03):
In some cases, a management letter may have been requested and is descriptive of sound practices or contains suggestions, but does not identify or recommend remedies for EAP practices that the accounting

firm believes are a departure from sound practices. In such a case, the
standard requires that the board of directors or a designated committee
reviews its contents.

IV.4.04 The EAP that reports less than $500,000 in revenues annu-
ally, and is not otherwise required to file an audit, undergoes
a review of financial statements, with a management letter if
applicable, and such a review is performed:

a. at the end of each fiscal year; and
b. by an independent certified public accountant approved
by the owners or board of directors.

IV.4.05 In an EAP that receives less than $500,000 in revenues and
does not conduct an audit, the owners or board of directors:

a. meets with the independent certified public accountant
within six months of completion to discuss the review of
the financial statements and any management letter that
may accompany this review; and
b. makes this review available for public inspection.

IV.5 *Payroll*

IV.5.01 The EAP reviews and approves payroll expenditures and:

a. documents changes in time and overtime records;
b. authorizes payment for new hires and severance for termi-
nations;
c. oversees mandatory deductions and rates of pay; and
d. ensures separation of payroll funds.

IV.5.02 The EAP's payroll practices comply with federal and state,
or provincial, wage and hour laws.

IV.6   Management of Investments

IV.6.01   In EAPs that invest funds, the owners/senior management or board of directors follows and biennially reviews an investment policy that outlines:

a. acceptable levels of risk;
b. criteria for contracting with investment advisors or firms; and
c. protocols for making investment decisions.

IV.6.02   A designated committee or agent of the owners/senior management or board of directors:

a. oversees and reviews both the investment of funds and the management, purchase, or sale of real estate, securities, and other assets;
b. ensures that practices conform to applicable legal and regulatory requirements; and
c. reports the status of investments and investment recommendations to the owners/senior management or board of directors.

V. EAP LEGAL LIABILITY

V.1   General Principles

V.1.01   The EAP utilizes legal counsel to clarify the meaning of laws or regulations governing the services it operates, or to respond to other legal inquiries.

Interpretation (V.1.01):
   The EAP consults with legal counsel regarding matters involving unusual disclosure of client information and associated risks, such as

when courts, public officials, investigative units, law enforcement bodies, or others request special or unusual information about an individual or family. This standard also applies to issues related to the confidentiality of records and the conditions under which they may be subpoenaed.

V.1.02   The EAP reduces its potential loss and liability by:

   a. assigning responsibilities related to liability matters to qualified persons whose job descriptions specifically include oversight of risk management; and
   b. developing a process to identify and analyze the nature, severity, and frequency of risks.

V.1.03   The EAP follows procedures regarding appropriate handling of media inquiries that protect the confidentiality of clients.

V.2   Liability Insurance

V.2.01   The EAP provides insurance coverage for staff members related to the scope of their activities performed on the EAP's behalf and:

   a. provides written notification to staff on the amount and type of such coverage;
   b. advises the staff of the extent and limits of such coverage; and
   c. the amount purchased is proportional to the EAP's size and nature of potential incidents.

Interpretation (V.2.01):
   All staff members must receive this information at the initiation of their association with the EAP and when any changes to the level and/or type of insurance coverage occur.

V.2.02   The EAP maintains appropriate insurance or bonding coverage for all EAP staff members who sign checks, handle cash or contributions, or manage funds to cover losses that may be incurred.

V.3   *Record-Keeping Practices and Procedures*

V.3.01   The EAP maintains client records in a manner consistent with applicable legal requirements and the EAP's confidentiality policy.

V.3.02   EAP policy states that a separate and distinct record is maintained for each client that is never part, of or stored with, any other record for the client.

Interpretation (V.3.02):
This standard includes instances when family members receive services as a group. Records are maintained on the individual. Other records include managed care, personnel, or medical records.

V.3.03   EAP policy establishes whether client records are the property of the EAP or the customer organization.

V.3.04   EAP record retention procedures address:

a. how long affiliates may retain client records;
b. whether affiliates are able to maintain a copy of the record after the record is returned to the EAP; and
c. whether affiliates are able to maintain a copy of the record after their contract with the EAP ceases.

V.3.05   After clients have ceased contact with the EAP, all client records are retained by the EAP for a minimum period of three years or as required by state or provincial law.

V.4 Security of Information

V.4.01　The EAP has procedures to protect service and organizational records, whether in electronic or paper form, from destruction by fire, water, loss, or other damage, and from unauthorized access, which include:

a. daily back-up of all electronic records;
b. electronic back-up maintained off-premises; and
c. storage of paper records in cabinets that are kept in a secure location, access to which is limited to those authorized to retrieve files in accord with X.2.02 and X.2.06.

Interpretation (V.4.01):
Secure storage may include locked file cabinets; a locked file room with limited access or a gatekeeper system whereby one person or a few people can unlock the file storage area or access the files themselves; and/or a system using a keypad or keys where only certain individuals are given the keypad code or copies of the keys. See also VII.8 Information Management.

V.4.02　EAP procedures govern the retention, maintenance, and destruction of records of former clients and include protocols on:

a. protection of privacy;
b. legitimate requests by former clients for access to information, when permissible by law;
c. requests for records of deceased clients; and
d. disposition of records in the event of the EAP's dissolution.

Interpretation (V.4.02):
Disposition of records may occur via use of a shredder or a destruction vendor.

V.4.03　The organization protects electronically maintained data as follows:

    a. all computers have up-to-date anti-virus protection; and
    b. secure protocols, including the use of passwords and fire-walls, govern the electronic collection and transfer of sensitive data.

V.4.04   The EAP that transmits information electronically complies with all applicable legal standards and requirements, including the use of appropriate formats, codes, and identifiers to ensure the security and privacy of such data.

V.4.05   To protect client files, access to databases, and confidential information, the EAP has security systems for programs in high-risk areas to deter facility break-ins after hours.

V.5   *Affiliate Agreements*

V.5.01   Affiliate agreements are comprehensive and address:

    a. the same performance standards required of EAP staff members, such as training and credentials; and
    b. roles and responsibilities of the EAP and affiliate.

Interpretation (V.5.01):
Please see section VIII.3 Competence of Affiliates for additional information on affiliate credentials.

V.5.02   Affiliate agreements include, but are not limited to:

    a. compliance with all applicable laws, including health, safety, and accessibility laws;
    b. record maintenance and destruction;
    c. access to records;
    d. transfer of confidential information;
    e. hours of operation; and
    f. use of standard data collection and client information forms.

Interpretation (V.5.02):

Record maintenance, as referenced in element (b) of this standard, must address the secure storage of client records to which only the affiliate has access.

V.5.03 Within the first three months of hire, all affiliates sign a statement acknowledging that they have received the following information:

a. the EAP's policies and procedures;
b. responsibilities related to mandated reporting, including identification of clinical indicators of suspected abuse and neglect, as required by federal and state or provincial law;
c. reportable criminal behavior, including criminal, acquaintance, and statutory rape; and
d. duty to warn.

V.5.04 Affiliate agreements state that EAP client records are the property of the EAP rather than the affiliate.

Interpretation (V.5.04):

Please see V.5.02 for additional information related to maintenance of EAP client records.

V.5.05 Affiliate agreements require the affiliate to keep records that reflect services provided for each session and the time and date of each session rendered.

V.5.06 Affiliate agreements stipulate that a minimum of five percent of the EAP's total number of affiliate records per year are evaluated on a semi-annual basis, or more frequently for affiliates with less than one year of experience with the EAP.

Interpretation (V.5.06):

Please see VII.6.02 for additional information on the annual site visit process for affiliates and VII.3.02 for the criteria to be used in this process.

V.5.07   Agreements with affiliates require them to carry professional liability insurance in the amount of $1 million/$1 million, or as required by state or provincial law.

V.6   Subcontractor Agreements

V.6.01   Agreements with subcontractors require the same quality and level of staff training as that of the EAP.

V.6.02   The EAP requires each subcontractor to keep records of all training/education provided to subcontractor staff and to make these available to the EAP and/or external reviewers upon request.

V.6.03   The EAP requires subcontractors to collect and report demographic information to the EAP on the clients they serve, including utilization rates, as defined by the EAP.

V.6.04   Subcontractor agreements stipulate that the EAP annually conducts random site reviews of a five percent sample of subcontractors to review the following:

    a. appropriateness of service delivery procedures;
    b. safety of physical facilities;
    c. possession of current licensure; and
    d. compliance with EAP contract requirements.

VI. CONTRACTS FOR EAP SERVICES

VI.1   Program Plans

VI.1.01   Each EAP contract includes a program plan that describes the facilities, equipment, and staff resources required.

VI.1.02 The program plan includes mechanisms for promotional and employee communications that include, but are not limited to, the following, as applicable:

a. printed communications;
b. company website;
c. referral resource database;
d. listserves, discussion groups, chat rooms, instant messenger, and other electronic communication tools;
e. training of supervisors, key management, and union representatives;
f. employee orientation; and
g. other promotional and educational activities.

VI.1.03 The EAP conducts a needs analysis, when requested by the customer organization, to facilitate program design, which includes:

a. a confidential survey of employee and management representatives to identify key problem areas;
b. employee profiles and demographics;
c. employee absenteeism rates;
d. employee turnover rates;
e. accidental injuries;
f. health insurance costs; and
g. worker's compensation claims.

VI.1.04 When legally permissible, and at the request of a customer, the EAP generates a profile that describes the work force demographics and the characteristics of the customer organization, which may include one or more of the following:

a. employee locations;
b. health coverage, including mental health benefits;
c. products or services provided by the customer organization; and

d. unionized or non-unionized setting.

VI.2   *Contractual Agreements*

VI.2.01   The EAP includes written information about its service design and implementation in all bids and proposals to customer organizations.

VI.2.02   The EAP establishes and abides by formal contractual agreements that include, but are not limited to:

a. objectives for the contract;
b. the services to be provided;
c. financial terms;
d. mutual indemnification, when appropriate; and
e. mutual obligations of the EAP and customer organization.

VI.2.03   Contractual agreements also specify:

a. evaluation methodologies;
b. quality improvement expectations;
c. administrative and clinical record-keeping procedures;
d. reporting procedures; and
e. auditing procedures.

VI.2.04   The EAP contract describes the professional qualifications of its staff members and affiliates.

VI.3   *Account Management Procedures*

VI.3.01   For each EAP contract, the EAP:

a. designates an account manager; and
b. delineates clear lines of responsibility for all aspects of the contract.

VI.3.02   For each EAP contract, the EAP also:

a. plans for implementation of the contract by the respective account manager and customer organization liaison; and

b. projects utilization rates.

VI.3.03   The EAP maintains up-to-date information on each customer organization's demographics, business, and covered EAP benefits.

VI.4   Contract Management with Customer Organizations

VI.4.01   Prior to initiating a contract, and annually thereafter, the EAP and customer organizations determine:

a. how a case is defined and how utilization is calculated;
b. how a "new" client is defined;
c. the amount (e.g., number of hours) of clinical and account management time projected per defined period;
d. desired outcomes and performance standards;
e. the means of measuring these outcomes; and
f. the format and frequency of reports.

Interpretation (VI.4.01):
The standard requires the EAP to designate the numerator and denominator for purposes of utilization as addressed in (a). Training to supervisors and other units are not acceptable factors to be addressed in utilization.

VI.4.02   The EAP retains copies of periodic reports on file.

VI.4.03   The EAP conducts an annual face-to-face or telephone interview with representatives of customer organizations to determine their satisfaction with the EAP and the service contract.

Interpretation (VI.4.03):

The surveys conducted as part of this standard may be included in the reports described in standard VI.5.04.

VI.5   Reports to Customer Organizations

VI.5.01   The EAP account manager reports to the customer organization at least quarterly, or as indicated by the EAP, on the demographics of clients served, including:

a. types of services requested;
b. number of sessions;
c. outcome measures, such as problem resolution rate and client satisfaction; and
d. any other indicators requested by the customer organization.

VI.5.02   The EAP provides customer organizations with utilization reports at least quarterly, or as indicated by the EAP, that address variables such as:

a. cases opened;
b. problem resolution rate;
c. website usage;
d. training/seminar participants;
e. client telephone contacts; and
f. supervisor telephone contacts, as applicable.

VI.5.03   Each customer organization receives a year-end summary that contains:

a. documented results of evaluation activities and whether the objectives for the EAP contract were achieved;
b. an explanation of successes or failures connected with each objective;

    c. assessments of the resources required/used to meet objectives;

    d. a plan for improving performance in areas needing improvement; and

    e. intended changes to the contract for the following year.

VI.5.04   The EAP provides the results of satisfaction surveys at least quarterly, or as indicated by the host or customer organization, and the EAP's owners, board of directors, and/or senior management.

Interpretation (VI.5.04):

The compilation of such data may be done independently or through a consultant.

VI.5.05   Reporting procedures include protections to guard against breaches of confidentiality for data requested by program managers of host or customer organizations.

VII. QUALITY IMPROVEMENT

VII.1   Quality Improvement Infrastructure

Interpretation (VII.1):

For the purposes of these standards, "long-term planning" and "strategic planning" are synonymous.

VII.1.01 The EAP quality improvement document:

    a. is annually reviewed and updated;

    b. is based on program objectives, contract expectations, the previous year's results, and services provided to customer organizations;

    c. includes a statement of goals and objectives for the coming year; and

    d. documents program evaluation methods that measure progress and results relative to these goals and objectives.

VII.1.02   The quality improvement document specifies:

    a. the information to be collected;

    b. procedures for retrieving and analyzing information;

    c. methods for monitoring and reporting results; and

    d. feedback mechanisms and corrective action.

Interpretation (VII.1.02):

Data collection mechanisms include, but are not limited to, client satisfaction questionnaires, organizational satisfaction questionnaires, personnel surveys, training evaluations, incident reports, and documentation of satisfaction at follow-up of service, as appropriate.

VII.1.03   At least every four years, the EAP conducts organization-wide, long-term planning which includes:

    a. reviewing the EAP's mission, values, and mandates, and making revisions, as necessary;

    b. establishing goals and objectives that flow from its mission and mandated responsibilities;

    c. assessing its strengths and weaknesses;

    d. assessing human resource needs; and

    e. identifying and formulating strategies for meeting identified goals.

VII.1.04   The EAP develops a short-term plan that supports the goals and objectives of its long-term plan and evaluates its progress toward achieving short-term goals and objectives at least quarterly.

VII.2   *Evaluation of Performance*

VII.2.01   The EAP evaluates its systems, procedures, and outcomes on an ongoing basis and uses the results to continuously improve performance.

Interpretation (VII.2.01):
  The EAP must take action based on the quality improvement findings to build on strengths, eliminate or reduce identified problems, determine possible causes when data reveal issues of concern, promulgate solutions and replicate good practice, and implement and monitor the effectiveness of corrective action plans, when indicated.

VII.2.02   The EAP collects and analyzes demographic data on clients and is able to generate customer profiles according to the following parameters:

  a. service provided;
  b. length of service;
  c. gender;
  d. age;
  e. job classification;
  f. religious affiliation, as appropriate to the EAP;
  g. acial/ethnic composition, as appropriate to the EAP; and
  h. amount the customer has paid per employee (for external programs only).

Interpretation (VII.2.02):
  The customer profile generated as per the requirements of this standard may be included in the reports described in standard VI.5.01.

VII.2.03   The EAP has a system in place to evaluate:

  a. the rate at which calls on the EAP's toll-free service access line are answered; and

   b. telephone call abandonment rates.

VII.2.04   The EAP evaluates the effectiveness of each host or customer organization's Drug Free Workplace program.

VII.2.05   An EAP that purchases training for supervisors, work-life, organizational development, or other services from subcontracting organizations has systems to monitor, evaluate, and improve each of those contracted services.

VII.2.06   The EAP evaluates referral resources on an on-going basis to assess the safety, quality, and effectiveness of services provided.

Interpretation (VII.2.06):
   Such evaluations may be conducted through site visits or investigations of the referral's reputation among its customers.

VII.3   Internal Quality Monitoring

VII.3.01   The EAP collects data on each service it provides and integrates findings into its overall quality improvement system, which may include, but not be limited to:

   a. training to supervisors and union representatives;
   b. work-life;
   c. critical incident stress management;
   d. legal services;
   e. short-term counseling; and
   f. assessment and referral.

VII.3.02   The EAP documents quarterly reviews of a statistically significant number of randomly selected open and recently closed client records, comprising at least five percent of the staff member cases per quarter, to evaluate:

a. the quality of assessments;
b. case or service planning;
c. services provided or obtained;
d. the results of the service; and
e. aftercare planning.

Interpretation (VII.3.02):
The client record review should be performed using a tool that specifies these and other indicators used.

VII.3.03 The quarterly internal client record reviews are:

a. conducted so that counselors and supervisors do not review cases in which they have been directly involved; and
b. distinct from periodic case review in which the counselor and supervisor review the client's progress towards achieving his/her service goals.

VII.3.04 The EAP annually reviews potential areas of liability and assesses areas of overall risk to the EAP, including, but not limited to:

a. research involving program participants, as applicable; and
b. compliance with legal requirements including licensing.

VII.3.05 At least quarterly, the EAP conducts a review of all grievances and incidents involving customer organizations and their employees, personnel, and affiliates, which includes a review of environmental risks.

Interpretation (VII.3.05):
Aggregated data on grievances and incidents are made available to the account managers and senior management or their designees.

VII.3.06   The EAP integrates the findings of external review processes including audits, accreditation activities, licensing, and other reviews into its quality improvement process.

VII.3.07   The EAP integrates the findings and recommendations of annual summaries and periodic reports developed for host or customer organizations into its quality improvement processes.

VII.4   External Audits

VII.4.01   The EAP complies with all host or customer organization requests for third-party audits as stipulated in contractual agreements.

VII.4.02   The EAP and host or customer organization mutually agree on the scope of third-party audits in advance, and audits are restricted to this predetermined scope.

VII.4.03   The third-party audit firm and/or its designated auditor are independent of the host or customer organization and the EAP.

VII.4.04   The EAP conforms to third-party auditing procedures when it permits components of files or records to be copied or removed from audited sites.

Note: Under United States federal regulations, bona fide audits, inspections, and evaluations are allowed without prior consent. Other federal legislation needs to be stated from other countries served. Records may be copied for audit purposes.

VII.4.05   A third-party audit firm signs non-disclosure and confidentiality agreements to protect the EAP's intellectual property.

VII.5   Outcomes Measurement

VII.5.01   In each of its programs, and on an ongoing basis, the EAP measures service outcomes for all clients, including:

   a. individual client satisfaction with all services received;
   b. level of functioning; and
   c. level of achievement for goal of assessed problem or request for service.

Interpretation (VII.5.01):
   All services include, but are not limited to, assessment and referral, information and referral, short-term counseling, CISM, and work-life. The compilation of such data may be done independently or through a consultant.

VII.5.02   EAP counselors give a satisfaction questionnaire to each client after the first session in all services.

VII.5.03   For its clinical services, the EAP collects and analyzes client self-reported outcomes and counselor-reported outcomes toward achieving the service goals established at the first session.

Interpretation (VII.5.03):
   According to this standard, goals may be addressed in the following domains: cognitive, behavioral, and emotional dimensions that enhance the client's ability to cope with occupational, social, psychological, and interpersonal problems.

VII.5.04   For its clinical services, the EAP uses standardized measurement tools at intake and at follow-up to evaluate improvements in client functioning.

Interpretation (VII.5.04):
   An example of such a measurement tool is the Global Assessment of Functioning (GAF) scale, which is available for clinical work in the United States.

VII.5.05   After each episode of training or consultation, the EAP surveys the service recipient to evaluate the impact of the training or consultation on his/her ability to handle a situation or situations in the workplace that the training or consultation was intended to address.

VII.6   Quality Improvement with Affiliates

VII.6.01   An EAP that purchases services from affiliates monitors and evaluates those contracted services and implements corrective action, if necessary.

VII.6.02   The EAP conducts random annual site reviews of a five percent sample of affiliates to review the following:

   a. appropriateness of clinical protocol and procedures, as addressed in VII.3.02;
   b. safety of physical facilities; and
   c. compliance with EAP contract requirements.

VII.6.03   In the five percent sample of affiliates reviewed in VII.6.02, the EAP also evaluates:

   a. attainment of the requisite credentials required of affiliates assuming different roles, e.g., assessment and referral, information and referral, and short-term counseling;
   b. possession of current licensure, certification, or registration; and
   c. attainment of ongoing training requirements set forth in the contract.

VII.7   Feedback Mechanisms

VII.7.01   At least annually, the EAP:

   a. shares findings from its quality improvement processes with personnel, clients, and other stakeholders; and

b. submits summary results of its planning and evaluation processes to the owners/senior management or board of directors, as applicable.

VII.7.02 Data from outcomes measurement and other quality improvement processes are distributed in a timeframe and form that are useful to all relevant staff.

VII.8   *Information Management*

VII.8.01 The EAP has a management information system that is capable of supporting its operations, planning, and evaluation activities.

VII.8.02 The EAP's management information system:

a. protects confidentiality; and
b. enables timely, and rapid access to information.

Interpretation (VII.8.02):
The management information system is capable of providing information without delay in emergency or crisis situations and within a timeframe that supports, rather than hinders, organizational decision-making and routine service-delivery functions.

VII.8.03 The EAP maintains a disaster recovery and back-up plan for all information systems that is designed to bring these services back into operation as quickly as possible following an interruption in service, and this system is tested at least once per year.

Interpretation (VII.8.03):
The EAP must maintain a back-up power supply or alternate system for phones and information management systems.

VII.8.04 The EAP has systems that maintain data capture methods for the following types of information:

a. client identification;
b. demographic and work data;
c. referral source;
d. presenting and assessed problems;
e. resolution of problems;
f. completion or termination of service;
g. other statistical data relevant to the quarterly and annual reports; and
h. outcomes measurement for services sought by clients.

VII.9   Corrective Action

VII.9.01   The EAP takes action based on the findings of its quality improvement processes to:

a. build on strengths;
b. eliminate or reduce identified problems;
c. determine possible causes when data reveal issues of concern;
d. promulgate solutions and replicate good practice; and
e. implement and monitor the effectiveness of corrective action plans, when indicated.

VII.9.02   The EAP revises policies and/or operational procedures, personnel assignments, personnel training, contracts, and services according to the recommendations of its quality improvement processes.

VIII. PERSONNEL AND AFFILIATE COMPETENCE

VIII.1   Competence of Counselors

VIII.1.01 EAP counselors have specialized training and demonstrated competence in all areas of EAP practice in which they are active.

VIII.1.02 Counselors possess knowledge and experience in the following areas:

a. family and relationship counseling;
b. addiction counseling;
c. short-term counseling, as applicable;
d. use of all clinical measurement tools, as appropriate to the service provided; and
e. instruction in assessment and referral.

VIII.1.03 All EAP counselors and their supervisors keep abreast of relevant regulatory and legislative developments.

VIII.2 Credential Requirements

VIII.2.01 The chief executive officer or his/her designee is qualified by:

a. an advanced degree from an accredited college or university;
b. at least five years of management experience;
c. assessed competence in administering and providing EAP services; and
d. management skills in addressing human resources and financial matters.

VIII.2.02 The EAP clinical director is qualified by:

a. an advanced degree from an accredited college or university in a field related to EAP services;
b. appropriate state or provincial licensure, certification, or registration;
c. at least two years of post-graduate experience in clinical services; and
d. assessed competence in administering and providing EAP services.

VIII.2.03 Assessment and referral, and short-term counseling personnel have at least the following qualifications:

    a. the terminal degree in a mental health profession (terminal degrees include an MSW, an MS in nursing, or a PhD in psychology);

    b. appropriate state or provincial licensure, certification, or registration; and

    c. training and experience in alcoholism/substance abuse treatment.

Interpretation (VIII.2.03):

This applies to services provided through all mediums, including face-to-face, telephone, and online services.

VIII.2.04 Assessment and referral, and short-term counseling personnel possess at least two of the following:

    a. training and experience in organizational dynamics;

    b. CEAP designation;

    c. at least 2,500 hours post-master's degree clinical experience;

    d. two years of EAP experience in a management or direct service role; or

    e. a completed master's level internship in an EAP setting.

VIII.2.05 Information and referral, and intake staff possess at least a master's degree in a mental health profession and at least one year of clinical practicum in social work, psychology, or another mental health profession.

VIII.2.06 EAP counselors who are working toward practice requirements for licensure or certification are supervised as per state, provincial, or professional licensure registration requirements.

VIII.3 Competence of Affiliates

VIII.3.01 All affiliates have specialized training and demonstrated competence in all areas of EAP practice in which they are active.

VIII.3.02 Criteria for selection of affiliates require:

   a. demonstrated competence in specialized areas, including those listed in VIII.1.02;
   b. ongoing attainment of state or provincially required professional development hours (PDHs), continuing education units (CEUs), or other professionally required training; and
   c. attainment of the training content addressed in section IX.4.

VIII.3.03 Affiliates that provide assessment and referral, and short-term counseling have at least the following qualifications:

   a. the terminal degree in a mental health profession (terminal degrees include an MSW, an MS in nursing, or a PhD in psychology);
   b. appropriate state or provincial licensure, certification, or registration; and
   c. training and experience in alcoholism/substance abuse treatment.

Interpretation (VIII.3.03):
   This applies to services provided through all mediums, including face-to-face, telephone, and online services.

VIII.3.04 Affiliates that provide assessment and referral, and short-term counseling possess at least two of the following:

   a. training and experience in organizational dynamics;

  b. CEAP designation;
  c. at least 2,500 hours post-master's degree clinical experience;
  d. two years of EAP experience in a management or direct service role; or
  e. a completed master's level internship in an EAP setting.

VIII.3.05 Affiliates that provide information and referral and intake services possess at least a master's degree in a mental health profession and at least one year of clinical practicum in social work, psychology, or another mental health profession.

VIII.3.06 In addition to the other requirements in this section, affiliates work in a clinical practice for a minimum of 10 hours per week.

VIII.4   Additional Credential Requirements

VIII.4.01 All EAP interns are in training at the undergraduate or graduate level and are closely supervised by licensed or certified practitioners.

VIII.4.02 Supervisors are qualified by additional training in supervision, and at least two years of supervised post-graduate experience in counseling.

Interpretation (VIII.4.02):
Supervisors that are administration or management specialists by training/education may have post-graduate experience that is non-clinical in nature.

VIII.4.03 Staff members and affiliates who provide management and human resource consultation possess a master's degree in a behavioral healthcare field with certification in human resource management and/or an approved EAP training program.

Interpretation (VIII.4.03):

For EAPs that use subcontractors or affiliates to provide work-life services, the EAP would include these qualifications as part of the contract and verify compliance through review of V.6.04 and VII.6.02.

VIII.4.04 EAP staff members and affiliates who provide Drug Free Workplace assessments possess master's degrees, licensure, and a minimum of three years' experience in a chemical dependency treatment program.

VIII.4.05 EAP staff members providing work-life services possess:

   a. a minimum of a bachelor's degree;
   b. specific expertise and training in the relevant field; and
   c. a minimum of 10 hours of training on appropriate intake and case management procedures.

Interpretation (VIII.4.05):

For EAPs that use subcontractors or affiliates to provide work-life services, the EAP would include these qualifications as part of the contract and verify compliance through review of V.6.04 and VII.6.02.

VIII.4.06 Work-life supervisors possess:

   a. a master's degree in a related field; or
   b. a bachelor's degree with a minimum of seven years' related experience.

Interpretation (VIII.4.06):

For EAPs that use subcontractors or affiliates to provide work-life services, the EAP would include these qualifications as part of the contract and verify compliance through review of V.6.04 and VII.6.02.

IX. STAFF SUPERVISION AND TRAINING

IX.1   Consultation with Staff and Affiliates

Interpretation (IX.1):

For purposes of these standards "supervision" can be used inter-changeably with "consultation" as it occurs with staff (versus affili-ates).

IX.1.01   EAP consultation emphasizes:

    a. clients' progress toward achieving goals and objectives;
    b. the safety and well-being of the client; and
    c. the quality of documentation in client records.

IX.1.02   Consultation is documented in the client's record and includes the supervisor's signature.

Interpretation (IX.1.02):

Supervision and/or consultation primarily address the elements out-lined in IX.1.01. Issues related to the quality of the work performed by staff and/or affiliates should be documented in the personnel or affiliate file or in separate files maintained by the supervisor, not in the client record.

IX.1.03   EAP staff and affiliates immediately report all critical incidents and cases which are potentially threatening to the client, the host or customer organization, or the EAP, and as part of consultation critical incidents are reviewed and information is documented.

IX.1.04   EAP procedures, or contract provisions with affiliates, address the frequency of consultation for staff members and affiliates.

IX.1.05 Consultation with EAP staff and affiliates occurs as follows:

    a. staff and affiliates who have less than three years of EAP experience have at least one hour of individual consultation for every 60 client contact hours;

    b. staff and affiliates with more than three years of experience have at least one half hour of consultation for every 90 client contact hours; and

    c. staff and affiliates in diverse geographical locations have at least one hour of consultation for every 90 client contact hours.

Interpretation (IX.1.05):

Consultation can be provided by telephone or through written or electronic contact.

IX.1.06 EAP staff members who provide work-life services receive a minimum of two hours of individual or group consultation per month.

Interpretation (IX.1.06):

For EAPs that use subcontractors or affiliates to provide work-life services, the EAP would include these qualifications as part of the contract and verify compliance through review of V.6.04 and VII.6.02.

IX.1.07 Consultation with staff and affiliates may occur electronically via the following mechanisms:

    a. email, provided that no identifying client data are exchanged; and

    b. website, provided that password protections exist, modes of transmission are secure, and the site communicates that confidential information is being transmitted.

IX.2   Supervision of Non-Clinical Personnel

IX.2.01   Face-to-face supervision of non-clinical staff occurs:

    a. at least quarterly for account managers and other non-clinical personnel, including off-site personnel; and

    b. at least monthly for telephone receptionists.

Interpretation (IX.2.01):

    The EAP monitors and evaluates how non-clinical staff members interact with clients, and provides them with suggestions for improvement during supervision.

IX.2.02   Supervision of account managers emphasizes the manager's ability to implement, monitor, and effectively coordinate EAP services.

IX.3   General Staff Training and Development Requirements

IX.3.01   All EAP personnel receive orientation to the EAP within 90 days of beginning work, and orientation is documented in the staff member's personnel record.

IX.3.02   The EAP orients all new personnel to:

    a. its mission, philosophy, and goals;

    b. its services, policies, and procedures;

    c. an organizational chart that delineates lines of accountability and authority at all levels of the EAP;

    d. the cultural and socioeconomic characteristics of the service population; and

    e. the EAP's relationship with other community resources.

IX.3.03   The EAP establishes a training and development program for counselors that is designed to:

a. improve their knowledge, skills, and abilities as they relate to the EAP field; and
b. be aware of, and sensitive to, the diverse backgrounds and needs of clients.

Interpretation (IX.3.03):

Training may address a variety of topics including: customer service, crisis intervention, brief therapy modalities, managed care, critical incident stress management, work-life programs, the impact of mental illness and substance abuse, work performance assessments, human resource management, culturally competent clinical practice, management information systems, legal practices, and other relevant topics.

IX.3.04 Each year, EAP counseling professionals complete required state or provincial professional development hours (PDHs), continuing education units (CEU), or their equivalent.

Interpretation (IX.3.04):

Such requirements are completed at accredited colleges/universities, other state or provincially licensed institutions, or at EASNA, EAPA, or CEAP-sponsored courses.

IX.3.05 Non-clinical staff complete a minimum of two hours of in-service training per year.

IX.3.06 Telephone receptionists receive training from the clinical and/or program director about the EAP's service delivery system.

IX.3.07 Training for staff members is documented and such documentation includes:

a. the attendee's name;
b. title of subject;

    c. number of hours;

    d. date of training; and

    e. the presenter's name and credentials.

IX.4   Training Content

IX.4.01   Counselors receive training within the first three months of employment on the following:

    a. EAP core technology and the optional EAP services;

    b. EAP theory and practice; and

    c. the application of counseling skills in a workplace setting.

Interpretation (IX.4.01):

In the event that training will not be offered for a time exceeding three months, the EAP must have a schedule of the upcoming trainings which staff will attend.

IX.4.02   At least annually, EAP counselors receive training on current issues related to addictions and crisis intervention.

Interpretation (IX.4.02):

Counselors must receive at least one hour of annual retraining on new issues in addiction treatment.

IX.4.03   Non-clinical administrative staff, such as account managers, receive training on the following:

    a. EAP models of service delivery;

    b. essential components of EAPs;

    c. prevention practices;

    d. outreach; and

    e. training techniques for managers, supervisors, and union representatives.

IX.4.04 EAP staff members receive training on ethical issues, client rights, and ways to protect those rights.

IX.4.05 Telephone and online counselors receive special training on non face-to-face counseling techniques before providing such services.

IX.5 *Risk Management Training*

IX.5.01 Personnel receive training on legal responsibilities regarding:

a. mandated reporting, including identification of clinical indicators of suspected abuse and neglect, as applicable;
b. reportable criminal behavior, including criminal, acquaintance, and statutory rape; and
c. duty to warn.

IX.5.02 The EAP annually trains personnel on emergency response practices, including:

a. the ability to assess risk and safety of clients;
b. techniques for handling emergencies; and
c. appropriate coordination with mental health, law enforcement, and other professionals.

Interpretation (IX.5.02):
The EAP should train personnel on all emergency response procedures described in III.4.01.

IX.5.03 The EAP trains all counselors on the following:

a. techniques for de-escalating conflict;
b. personnel safety measures;
c. management of aggressive or out-of-control behavior; and

    d. protocols for notifying family members, legal guardians, or other contacts in the case of emergencies.

X. PROFESSIONAL PRACTICE

X.1 Protection of Rights

X.1.01   All clients receive summary information about their rights and responsibilities through a Statement of Understanding that is:

    a. provided in writing;
    b. distributed and explained at the beginning of his/her initial appointment;
    c. available in the major languages of clients at host or customer organizations, including English;
    d. effectively and appropriately communicated to persons with special needs; and
    e. signed by the client.

Interpretation (X.1.01):
For clients receiving telephone services, the EAP staff member should read the Statement of Understanding prior to service delivery to obtain the verbal confirmation of the client's understanding regarding his/her rights and responsibilities, and document this in the client's progress notes. Clients receiving online services should provide an electronic confirmation of their understanding.

X.1.02   The Statement of Understanding includes the following:

    a. eligibility criteria;
    b. financial terms;
    c. limitations to the EAP's confidentiality obligations; and
    d. the client's legal rights regarding EAP service use.

Interpretation (X.1.02):

Please reference section X.4 "Confidentiality and Privacy Protections for Clients" in relation to (c) limitations to the EAP's confidentiality obligations.

X.1.03   The Statement of Understanding also describes the client's rights and responsibilities including, but not limited to:

    a. basic expectations for use of the EAP's services;
    b. hours during which services are available; and
    c. the inclusion of clients, or as appropriate, his/her parent or legal guardian, in decisions regarding the services provided.

X.1.04   The EAP follows a policy that addresses the client's right to receive services in a way that does not stigmatize him/her or jeopardize his/her employment.

X.1.05   All clients have the right to be treated equitably and without favoritism, subject to limitations imposed by contractual obligations.

X.1.06   The EAP informs each client that if s/he uses a company computer, information transmitted may be tracked by the host or customer organization.

Interpretation (X.1.06):

This standard includes company computers used at the client's home.

X.2   Access to Files and Records

X.2.01   The EAP informs each client at the beginning of service that a record is kept documenting all service contacts, including the date and time of each occurrence, and services provided.

X.2.02   Record access is limited to:

    a. the client, or as appropriate, his/her parent or legal guardian;

    b. staff and affiliates authorized to see specific records on a "need to know" basis; and

    c. others outside of the EAP whose access to the information in the record is permitted by law or granted by the client.

Interpretation (X.2.02):

Records should not be left in public areas such as on carts in hallways, on desks, or in insecure areas. When not being used by authorized staff, files should be returned to a secure area.

X.2.03   Clients or their designated legal representatives receive instruction on how to access client records and such procedures are consistent with applicable rules, regulations, and laws, and the EAP's professional judgment as to the clients' best interest.

X.2.04   If the EAP determines that allowing a client to review his/her record would be harmful, and applicable law neither prohibits nor requires direct record access by the client, then:

    a. senior management reviews, approves in writing, and enters into the client record the reasons for refusing to allow a client to review his/her record; and

    b. the EAP has procedures that permit a qualified professional to review the records on behalf of the client, provided that the professional signs a written statement that the information determined to be harmful will not be provided to the client.

Interpretation (X.2.04):

A person's right to review his/her care or treatment may be denied or otherwise limited only in the most extreme circumstances where serious harm is likely to ensue. In such cases, objective criteria must guide deci-

sions to deny access. In all cases, the EAP must operate in accord with applicable law.

X.2.05   Clients have the right to insert a statement into their records, and if personnel insert a statement in response, such statements are inserted with the knowledge of the client, and the client is given the opportunity to review such a response.

X.2.06   EAP administrators and staff, including auditors and third-party evaluators, are permitted access to EAP records for the purposes of:

a. program oversight, evaluation, and quality improvement;
b. destroying EAP records at the end of their period of maintenance; and
c. transferring EAP records from one affiliate to another.

Interpretation (X.2.06):
Confidentiality must be maintained during this process.

X.3   Grievance Procedures

X.3.01   Procedures provide clients with a formal mechanism for expressing and resolving complaints and grievances, and such procedures:

a. are given to all clients at the time of intake and upon request, or at the initiation of a grievance;
b. include an appeals procedure;
c. provide for a timely resolution of the matter; and
d. require a written response to the aggrieved that includes documentation of the response in the client record.

X.3.02   Complaint mechanisms include procedures to inform the EAP's management and board of directors, if applicable, of problems with service provision or other matters of concern.

X.3.03   The EAP makes summary data regarding complaints available to external auditors or reviewers, as requested.

X.4   Confidentiality and Privacy Protections for Clients

X.4.01   The EAP gives clients written information that describes the EAP's confidentiality procedures, and requires clients to sign a statement indicating their understanding of their confidentiality rights and any limitations.

X.4.02   Staff members, affiliates, consultants, auditors, temporary staff members, and student interns sign a confidentiality agreement specifying that they agree to uphold the EAP's confidentiality practices inside and outside of the EAP offices.

X.4.03   EAP procedures governing access to, use of, and release or disclosure of information about clients meet applicable legal requirements under federal and state or provincial law.

X.4.04   External EAPs use a billing format that protects client confidentiality.

X.5   Releases

X.5.01   The EAP obtains written consent to release information from a client when:

a. s/he is referred out of the EAP to enable the counselor to follow-up to assure the referral has been completed;
b. s/he must be absent from work to participate in treatment or to be hospitalized; or
c. there are any other circumstances that require communication by the EAP regarding a client's confidential information.

Interpretation (X.5.01):

All releases must meet requirements set forth in federal and state or provincial laws.

X.5.02 The EAP assumes a protective role regarding the disclosure of information about clients and has clearly stated procedures governing the disclosure of such information, including instances where the client may be dangerous to him/herself or others.

Interpretation (X.5.02):

The EAP's procedures must reconcile legal restrictions and requirements on the release of identifying information about clients with mandatory reporting requirements and the EAP's duty to warn a person who may be in danger. Procedures should include guidance to personnel for reporting a person who is suspected to be a danger to him/herself or the community.

X.5.03 When the EAP receives a request for the release of confidential information about a client, or the release of confidential information is necessary for the provision of services, prior to releasing such information, the EAP:

    a. determines if the request is valid and in the best interest of the client;

    b. obtains the informed, written consent of the client; and

    c. if the person is an adult or minor who is incapable of providing informed consent, obtains consent from his/her parent or legal guardian.

Interpretation (X.5.03):

In the context of this standard, "valid" means justifiable, legitimate, convincing, legally permissible, and in the best interest of the client. Information should be released on a "need to know" or "need to access" basis and should be limited to portions of documents needed to answer a business, legal, or other legitimate inquiry. Consent is not necessary where the request is pursuant to a court order, audit, duty to warn, or emergency situation.

X.5.04   Informed, written consent includes the following elements:

     a. the signature of the person whose information will be released, or the parent or legal guardian of a person who is unable to provide informed consent;
     b. the specific information to be released;
     c. the purpose for which the information is to be used, except where disclosure is mandated by law or the client is receiving service under court supervision or directive;
     d. the date the consent takes effect;
     e. the date that the consent expires, not to exceed 90 days from the date consent is given for a one time release of information, or one year or as otherwise required by law when the release of information is required for ongoing service provision by a contracted or cooperating service provider;
     f. the name of the person to whom the information is to be given;
     g. the name of the EAP staff person who is providing the confidential information; and
     h. a statement that the client may withdraw his/her consent at any time.

Interpretation (X.5.04):
   Blanket consent forms signed by clients when service is initiated do not meet the requirements of this standard.

X.5.05   The EAP's confidentiality procedures address the release of information for worker's compensation and disability claims.

X.6 *Conduct of Staff and Affiliates*

X.6.01 The EAP adopts and follows its own code of ethics, requires its professional staff and affiliates to adhere to the codes of ethics of their respective professions, and avoids conflicts of interest in carrying out its responsibilities.

X.6.02 The EAP provides customer organizations with a copy of relevant procedures related to its code of ethics which:

a. establish protections against conflict of interest in making referrals; and
b. prohibit making or accepting payment or other consideration in exchange for referrals.

Interpretation (X.6.02):
A corporate conflict of interest policy may be used to meet this requirement, if such a procedure applies to EAP professionals.

X.6.03 If an EAP contract permits a client to continue to receive services from his/her EAP counselor following completion of contracted services, the following conditions apply:

a. the contract clearly indicates that no self-referrals will be made to an EAP counselor before the allotment of EAP sessions has been used;
b. the client is given a referral choice of at least two different organizations or individuals, one of whom may be the current EAP counselor; and
c. the client signs a freedom of choice understanding, which is maintained in the client's record.

X.7 *Conflicts of Interest*

X.7.01 In not-for-profit and publicly traded for-profit EAPs, the EAP's conflict of interest policy mandates that members of the board of directors and personnel who are involved individually, or as part of a business or professional firm, in the EAP's business transactions, leases, or current professional services, or who have a financial interest in the organization's assets, disclose this relationship and do not participate in any discussion or vote taken with respect to such transactions, services, or interests.

X.7.02   In not-for-profit and publicly traded EAPs, the conflict of interest policy prohibits preferential treatment of members of the board of directors, personnel, or contractors in applying for and receiving the EAP's services.

X.8   Research Policies and Procedures

X.8.01   EAP policy clearly states whether or not it conducts, participates in, or permits research involving clients.

X.8.02   The EAP has a mechanism, such as a human subjects committee or an internal review board, that reports to the chief executive officer or his/her designee, or the board of directors, and:

   a. reviews research proposals that involve clients;
   b. makes recommendations regarding the ethics of proposed or existing research;
   c. makes recommendations as to whether or not to approve research proposals; and
   d. monitors ongoing research activities.

X.8.03   Participation in research is voluntary, and the EAP:

   a. does not threaten to withdraw services or otherwise coerce clients into participating; and
   b. follows procedures governing the use of modest incentives for attracting and retaining participants, as applicable.

X.8.04 Each research participant, or his/her parent or legal guardian, signs a consent form that includes:

    a. a statement that s/he voluntarily agrees to participate;
    b. a statement that the EAP will continue to provide services whether or not s/he agrees to participate;
    c. an explanation of the nature and purpose of the research;
    d. a clear description of possible risks or discomfort;
    e. a guarantee of confidentiality; and
    f. the participant's signature.

X.8.05 The EAP safeguards the identity and privacy of clients in all phases of research conducted by or with the cooperation of the EAP.

Interpretation (X.8.05):

Statistical analyses, reports, and summaries are compiled and presented in a manner that masks the identity of the client. Case examples and extracts from individual records must be prepared, prior to dissemination, in a manner that masks the individual's identity.

X.9 *Ethical Considerations Related to Web-Based Services*

X.9.01 The EAP that delivers services via a website and requires clients to provide identifying information and/or tracks individual client use of the website, establishes a privacy policy that:

    a. is clearly posted on the website;
    b. protects the confidentiality of website users;
    c. advises clients to read the privacy policy of any linked sites before providing identifying information to such sites; and
    d. prohibits the EAP from selling, or otherwise making user data available to third-parties.

X.9.02   The EAP further protects client privacy by using up-to-date security technology, including firewalls and encryption software.

X.9.03   All website content is reviewed by qualified professionals prior to its inclusion on the EAP's website.

X.9.04   The EAP establishes timeframes that specify how often the website is updated to ensure that information is current, reliable, and free from error.

X.9.05   The EAP website provides clients with information on the following:

a. how to contact a live, qualified EAP counselor or affiliate; and
b. phone numbers to use in case of an emergency.

Interpretation (X.9.05):
This information should appear on each page of the EAP's website.

X.10   *Ethical Considerations Related to the Use of Specific Technologies*

X.10.01   An EAP that provides services online and/or via telephone adheres to all applicable ethical and legal standards and requirements related to such modes of service.

Interpretation (X.10.01):
In cases where an EAP provides online and/or telephone services across state lines, the EAP must comply with the licensure requirements of the state in which the client resides.

X.10.02   When client information is sent or transmitted, the EAP:

  a. includes a statement that the information is confidential, and that the receiver may not re-disclose the information without permission; and

  b. verifies receipt of the information.

Interpretation (X.10.02):
  Client information is not communicated via facsimile except in urgent circumstances.

X.10.03   An EAP that provides services via telephone informs the client when the call is being monitored for any reason.

XI. INTAKE, ASSESSMENT, AND SERVICE PLANNING

XI.1   Access Procedures

XI.1.01   Procedures for accessing EAP services:

  a. minimize barriers to the timely initiation of services or use of services; and

  b. give priority to employees or eligible participants with urgent needs or in emergency situations.

XI.1.02   The EAP communicates to customers, employees, and eligible participants that access to the EAP's services occurs through one of the following mechanisms:

  a. self-referral by employees and eligible participants for problems that may be adversely affecting their job performance;

  b. referrals by supervisors and suggestions by union representatives, human resources, and/or medical personnel; and

  c. mandatory referrals.

XI.1.03   The EAP direct service staff have immediate access to information on the benefits conferred to employees under the terms of each contract.

Interpretation (XI.1.03):
Such benefits may include, for example, utilization requirements or maximum number of sessions.

XI.1.04   Regardless of the type of service, the EAP follows procedures for dealing with client problems that occur during and outside of work hours, which include the following:

a. life threatening emergency situations are addressed immediately, 24 hours a day, seven days a week, 365 days a year;
b. counselors with clinical backgrounds are available by telephone to respond to emergencies and are able to access appropriate resources, either directly or by referral; and
c. non-life threatening emergencies are addressed by the end of the next business day.

XI.1.05   EAP services provide toll-free and/or online access, as appropriate, 24 hours a day, seven days a week, 365 days a year to:

a. employees and eligible participants needing help; and
b. host or customer organization representatives seeking assistance with organizational problems such as traumatic work site incidents.

XI.1.06   Employees and eligible participants are able to see counselors in person before, during, or after work hours, and counselors:

a. are available within 30 miles of client homes or work sites, unless the geography of the area prohibits such availability;

b. are located near public transportation;

c. offer appointments at least one evening a week; and

d. provide clear directions to the counseling site.

XI.1.07 The EAP adjusts its staffing patterns and availability to accommodate the working hours of employees at the host or customer organization.

XI.2 *Intake Process*

XI.2.01 The EAP follows standard procedures for intake in all services, including face-to-face counseling, telephone, and online services, as applicable.

XI.2.02 At the point of intake, intake personnel immediately establish if the client is experiencing an emergency and provide services, as appropriate.

XI.2.03 At intake, the EAP provides the following in writing, or explains telephonically with follow-up documentation:

a. referral resources and referral procedures;

b. the service and the number of sessions or contacts based on contractual agreement; and

c. follow-up procedures.

XI.2.04 The EAP bases its intake methods on the services it provides and the needs of its clients.

XI.3 *General Assessment Requirements*

XI.3.01 At the intake interview or initial assessment, EAP counselors obtain relevant assessment information including, but not limited to:

  a. demographic information;
  b. the nature of the request or presenting problem;
  c. any work-related issues; and
  d. a diagnosis, as applicable and when required by the EAP contract.

XI.3.02   The assessment includes information used to determine whether to retain a case for short-term counseling, or whether it is appropriate to refer the case to outside or approved resources for ongoing treatment.

Interpretation (XI.3.02):

  Services that require referral include, but are not limited to, presenting problems that involve psychosis, inpatient needs, addiction treatment, or severe medical complications.

XI.3.03   At the initial counseling session, the counselor and the client:

  a. assess the underlying problem and complete the assessment;
  b. determine the goals to be achieved; and
  c. develop the preliminary service plan.

XI.4   *Clinical Assessments*

Note: The term "clinical" in these standards may be used interchangeably with the term "biopsychosocial."

XI.4.01   Clinical assessments contain, at a minimum, the following information:

  a. environment and home situation;
  b. religion, if appropriate;
  c. financial status and health insurance, if appropriate;
  d. social and peer groups;

e. interests, skills, and aptitudes;

f. work history and military service, as applicable;

g. education; and

h. date of last medical exam.

XI.4.02 Clinical assessments also include:

a. physical illness/somatic variables/medical treatment;

b. the use of alcohol or other drugs;

c. behavioral/cognitive patterns that cause health risks, based on physical, emotional, behavioral, or social conditions; and

d. when appropriate, legal, vocational, and/or nutritional needs of the client.

XI.5 *Referrals*

XI.5.01 The EAP immediately notifies clients if it cannot promptly provide needed services.

XI.5.02 The EAP has procedures to facilitate client referrals which address the provision of consultation between the EAP and the host or customer organization, and responsibilities for providing follow-up, aftercare, and transition for clients.

XI.5.03 The EAP makes a referral when the client requires services beyond the stated or contractual mandate of the EAP, or when specialty resources are not available through the EAP.

Interpretation (XI.5.03):

For example, in the case of a short-term counseling model it may be appropriate to refer the client for alcohol/drug rehabilitation or psychiatric care.

XI.5.04 When making referrals, the EAP informs clients that they will be responsible for the cost of services beyond those provided by the EAP, and/or of any liabilities that may be incurred, such as insurance company access to files.

XI.5.05   The EAP uses the client's unique needs to guide referrals to organizations or any providers of goods and services.

Interpretation (XI.5.05):
The client's "unique needs" include level of care, geography, clinical requirements, or preference for a provider of a particular gender or ethnic background, when doing so is legally permissible.

XI.5.06   The EAP maintains a log of clients and referral sources for persons who are referred out of the EAP for services, which includes the following:

    a. specific referral source;
    b. date and method of referral;
    c. date the case is opened;
    d. date of the first appointment offered;
    e. date of the first face-to-face appointment;
    f. client identifier;
    g. presenting problem;
    h. disposition;
    i. follow-up schedule (to be included only in the client record);
    j. name of counselor making the referral; and
    k. date the case is closed.

XI.5.07   The case manager or EAP counselor conducts follow-up on referrals to outside agencies and documents the results for evaluation purposes.

XI.6   Outreach

XI.6.01   Mechanisms for providing information to employees and eligible participants include, but are not limited to:

    a. brochures, newsletters, and flyers;

    b. introductory letters;

    c. orientation sessions;

    d. supervisory and/or key employee training;

    e. promotional activities;

    f. educational activities; and

    g. internet and intranet communications.

XI.6.02   The EAP offers targeted promotional material to eligible participants every 18 months, or as directed by the host or customer organization.

XI.7   Special Service Delivery Considerations

XI.7.01   When the client is a victim of abuse or neglect, the EAP intervenes with more intensive services and provides more frequent monitoring and coordination with providers.

XI.7.02   EAP procedures address back-up and support in managing cases that involve threats of violence, including homicidal or suicidal ideation.

XI.7.03   The EAP provides culturally diverse and accessible services that include:

    a. assumption of service responsibility in rural and remote areas;

    b. assumption of service responsibility for persons with visual, hearing, and physical impairments; and

    c. diversity training for all professional staff associated with the program.

XI.7.04   Service planning and delivery meet the diverse and unique needs and preferences of clients.

Interpretation (XI.7.04):

Diverse and unique needs and preferences may be related to age, sex, gender, sexual orientation, physical limitations, ethnicity, culture, and other characteristics.

XI.8   EAP Staffing Patterns and Ratios

XI.8.01   The EAP maintains staff ratios for short-term counseling services as follows:

   a. for eight session models, one full-time equivalent counselor or affiliate to 3,500-4,000 employees; and
   b. for six-session models, one full-time equivalent counselor or affiliate to 4,000-5,000 employees.

Interpretation (XI.8.01):

These ratios assume a 45-minute, telephone or face-to-face session, with a 5% utilization rate.

Note: When the employee population is located over a large geographical area, the ratio is determined by access to the counseling location, with the overriding criterion being the availability of a counselor within 30 miles of the client's worksite or home. EAP administrative and support staff should not be included in the staff-to-eligible population ratio unless administrators handle cases on a regular basis. At no time is the chief executive officer or equivalent to be considered more than a half-time counselor.

XI.8.02   An EAP that provides information and referral, or assessment and referral, maintains at least one counselor or affiliate to 8,000 employees.

Note: When the employee population is located over a large geographical area, the ratio is determined by access to the counseling location, with the overriding criterion being the availability of a counselor within 30 miles of the client's worksite or home. EAP administrative and support staff should not be included in the staff-to-eligible population ratio unless administrators handle cases on a regular basis. At no time is the chief executive officer or equivalent to be considered more than a half-time counselor.

XI.8.03   EAP counselors that provide short-term counseling average no more than 28 hours per week of counseling over any four-week period.

Interpretation (XI.8.03):
Counselors may be full-time or full-time equivalents. The standard applies whether counseling is provided face-to-face or via telephone.

XI.8.04   The EAP maintains a formal relationship with a board-certified psychiatrist who provides on-call consultation.

XI.9   Client Records

Note: All policies and procedures pertaining to client records apply to subcontractors and affiliates, as well as staff offices.

XI.9.01   The EAP maintains a record that addresses EAP service use for each client.

XI.9.02   Client record entries contain only the information that is necessary to properly serve the client.

XI.9.03   The EAP client record contains:

    a. a statement of the presenting problem, as appropriate to the service provided;
    b. demographic information on the client, including age, sex, and ethnicity;

c. results of any assessments;

d. service plans; and

e. progress notes.

XI.9.04   The client record also contains:

a. a detailed account of the supervision or consultation, including data for the recommendations and actions taken;

b. follow-up plans; and

c. a closing summary.

Interpretation (XI.9.04):

Supervision and/or consultation primarily address the elements outlined in IX.1.01. Issues related to the quality of the work performed by staff and/or affiliates should be documented in the personnel or affiliate file, or in separate files maintained by the supervisor, not in the client record.

XI.9.05   When necessary due to the nature of individual needs and/or the type of service being provided, basic information is supplemented by psychological, medical, or biopsychosocial evaluations.

XI.9.06   Upon termination of service, or within 30 days of termination, a closing summary is entered into the client record which includes:

a. a report of changes in condition regarding the assessed problem;

b. recommendations for further action by the client and employer; and

c. referral or recommendations for any future services, as appropriate.

XI.9.07   All record entries for services are completed, signed, and dated by the person who provided the service.

XI.9.08   The EAP screens its client records for unsummarized notes, observations, and impressions, and other material that should be expunged at the closing of the record, and at least annually.

Interpretation (XI.9.08):
This standard applies to both hard copy and electronic records.

XII. SERVICE DELIVERY

XII.1   Prevention Services

XII.1.01   The EAP provides prevention services that address the following components:

a. outreach;
b. health promotion and wellness; and
c. coordination with healthcare providers.

XII.1.02   Topics addressed in prevention activities are changed and updated to reflect the needs and feedback of the host or customer organization and its employees.

XII.1.03   The EAP emphasizes the importance of prevention in all of its activities and offers to provide at least one primary prevention activity annually, for the host or customer organization.

XII.1.04   The EAP develops and offers educational sessions on wellness and other prevention-related topics.

XII.2   Training of Supervisors and Union Representatives

XII.2.01   The EAP provides general education and training to supervisors, union representatives, human resource professionals, safety committee members, benefits managers, and other key employees, as applicable.

XII.2.02   Supervisory training addresses how to recognize signs of deteriorating job performance and the proper means of documenting this in the personnel record.

XII.2.03   The EAP provides individual supervisors and human resource professionals with training on how to make referrals to the EAP for management of employee job performance and behavioral problems.

XII.2.04   Within four months of program start-up, the EAP, at the discretion of the host or customer organization, provides training which includes, but is not limited to:

a. the philosophy of the EAP;
b. confidentiality procedures and protections;
c. range of services provided;
d. location of offices, hours of operation, and telephone numbers; and
e. roles and responsibilities of management, supervisors, and union representatives, as applicable.

XII.3   Organizational Development

XII.3.01   The EAP offers organizational development services on the following topics:

a. needs assessment;
b. policy development;
c. team building; and
d. executive coaching.

XII.3.02   The EAP offers education and management consultation to host or customer organizations on other matters including:

    a. critical incident stress management protocols;

    b. managing change;

    c. smoke free workplace;

    d. workplace violence; and

    e. managed care.

XII.3.03 Ongoing management consultation is available to host or customer organizations as they help clients reintegrate into the workplace.

Interpretation (XII.3.03):

Consultation is provided within the confidentiality provisions and with a signed release of information from the client, and adherence to reasonable accommodation laws.

XII.3.04 The EAP includes a disclaimer in all contracts which states that any advice or recommendations made during the course of providing consultation to management is not and cannot be construed as a legal opinion.

XII.3.05 The EAP includes disclaimers in its promotional material to clarify that opinions expressed are informational and are not meant to represent the host or customer organization or the EAP.

XII.4 Critical Incident Stress Management

XII.4.01 The EAP provides critical incident stress management (CISM) and supportive services when a host or customer organization faces a crisis situation.

XII.4.02 Each EAP contract includes a clear definition of a critical incident.

XII.4.03 The EAP consults with a designated representative of the host or customer organization to determine if CISM is an appropriate intervention.

XII.4.04   The EAP trains its CISM counselors to assess the level of intervention that is required in particular situations.

XII.4.05   For CISM services, the EAP collects and analyzes data on the following for host or customer organizations:

a. size of groups;
b. number of sessions and/or number of hours spent in total; and
c. number of counselors required per incident.

XII.5   *Drug Free Workplace Services*

XII.5.01   The EAP offers a needs assessment to determine:

a. what components of a Drug Free Workplace are most appropriate for the host or customer organization; and
b. for which of the identified components the EAP will be providing services.

XII.5.02   Upon request, the EAP provides the host or customer organization with samples of Drug Free Workplace policies and procedures that meet federal and state guidelines.

Interpretation (XII.5.02):
   Samples of Drug Free Workplace policies may include, but are not limited to, information on drug testing procedures, drugs being tested for, who can be tested and under what conditions, the consequences of refusing to take a test, testing positive, and who is responsible for paying for the drug test.

XII.5.03   The EAP assists in the development of an implementation schedule of Drug Free Workplace requirements and provides training sessions on implementation which include:

a. the supervisor's responsibility in implementing the policy;
b. information related to Department of Transportation regulations, and/or the Drug Free Workplace Act;
c. how to deal with performance problems when personal problems are a contributing factor;
d. the expectation that a supervisor-initiated referral is based solely upon performance issues or a request for assistance by the employee; and
e. procedures to deal with referrals resulting from confirmed drug tests.

Interpretation (XII.5.03):
The procedures described in (b) refer to United States legislation and programs, and do not apply in Canada nor in other countries.

XII.5.04 The EAP provides employee education that addresses the following:

a. the Drug Free Workplace policy, including how to understand, cooperate with, and benefit from the policy and program;
b. the types of assistance available through the EAP; and
c. information about substance abuse and the dangers of substance use on the job.

XII.5.05 Training for supervisors includes how to:

a. observe and assess employee job performance over time;
b. recognize and assist employees who have job performance problems that may be associated with alcohol or drugs;
c. refer an employee for assistance;
d. document employee successes and problems;

     e. enforce the Drug Free Workplace policy consistently; and

     f. maintain confidentiality.

Interpretation (XII.5.05):

When drug testing is a part of the program, training must also include information on testing policies and procedures.

XII.5.06 When drug/alcohol testing is part of the Drug Free Workplace program, such testing is administered in a manner that protects the anonymity and confidentiality of self-referred clients.

XII.5.07 The EAP offers a comprehensive assessment, referral, and monitoring program for employees who test positive for, or self-identify as, drug or alcohol abusers, and procedures for such a program include:

     a. a Statement of Understanding;

     b. the differences between the Drug Free Workplace program and other EAP services;

     c. the limits of confidentiality;

     d. a standardized chemical dependency assessment form;

     e. the standardized level of care criteria; and

     f. a standardized Release of Information form for collateral contacts.

XII.5.08 When the EAP provides Substance Abuse Professional services, it complies with Department of Transportation regulations and other federal and state regulations.

XII.6   *Work-Life Services*

XII.6.01 The EAP conducts a needs assessment of the host or customer organization to determine the most appropriate and effective work-life services for the host or customer organization and its employees.

XII.6.02 EAP assessment procedures include the use of a work-life intake tool to evaluate client needs.

XII.6.03 Materials distributed as part of work-life services are:

a. accurate and well-researched;
b. non-discriminatory; and
c. updated at least every three years.

Interpretation (XII.6.03):
This includes work-life information provided via the EAP's website.

XII.6.04 The EAP maintains up-to-date information on referral resources which includes, but is not limited to, the following:

a. contact information;
b. type of service offered;
c. licensure information;
d. location; and
e. cost for clients.

XII.6.05 The EAP annually conducts random site reviews of five percent of its referral resources to assess the following:

a. quality of the service provided;
b. safety and accessibility of the physical facilities; and
c. possession of current licensure, as applicable.

Interpretation (XII.6.05):
Agreements with subcontractors that provide work-life information and referral require the same review of referral resources as that done by the EAP.

XII.6.06   The EAP keeps a record of clients who request work-life services.

XII.7   Work-Life: Legal Services

XII.7.01   Contracts with the legal services providers address whether self-referral for associated legal services is permitted.

XII.7.02   All attorneys used by the EAP have expertise in the area in which they practice or self-refer.

XII.7.03   The EAP informs clients of the types of legal services that are covered by the EAP, such as legal advice services other than representation, and the number of hours allowed.

XII.7.04   The client is informed of all fees for legal services during the initial consultation session, and fees for self-referrals for legal services are reasonable.

XII.7.05   The EAP has a procedure that requires legal services providers to make referrals only to attorneys who practice in the state in which the client's legal matter resides.

XII.7.06   If the legal services provider identifies other issues relevant to the client, s/he must refer the client back to the EAP, as appropriate.

Interpretation (XII.7.06):
   Other issues may include, but are not limited to, co-existing mental health or family concerns, such as stress from a divorce or child custody arrangements.

XII.7.07   If the host or customer organization and the EAP permit self-referral for legal services, the legal services provider offers the client a choice by providing the name of at least one other attorney who is separate from any firm with which the EAP may have a financial interest.

XII.7.08 The EAP requires all legal services providers to report the following information, at least quarterly:

a. number of cases;
b. case disposition;
c. type of cases; and
d. number of cases referred back to the EAP.

XII.8 Information and Referral, and Assessment and Referral Services

XII.8.01 The EAP can demonstrate a rapid and effective response in linking clients in need with appropriate EAP resources and supportive interventions.

XII.8.02 The EAP maintains records of referrals made to collaborating organizations and conducts routine follow-up on a sample of referred cases to determine whether individuals receive the services for which they were referred.

XII.8.03 When the EAP's information and referral service has a particular focus, such as elder care or legal services, the EAP adequately informs the participants of this specialization.

XII.8.04 For each service offered to the host or customer organization, the EAP maintains separate policies and procedures addressing eligibility, access, financial terms, and other essential issues.

XII.9 Short-Term Counseling

XII.9.01 The EAP provides clinical services through a comprehensive, formal delivery system.

XII.9.02   For clients with mental health and/or substance abuse problems, qualified counselors document, as permitted by law, the results of their assessments which may include a diagnostic summary, and/or a diagnosis according to the current Diagnostic and Statistical Manual of Mental Disorders of the American Psychiatric Association, or relevant assessment criteria of the respective country.

XII.9.03   EAP counselors establish brief service goals that are behaviorally focused and measurable, and such goals are based on the assessed problem(s).

XII.9.04   The service plan builds on strengths, engages clients in resolving their problems or requests, and is focused on the timely resolution of the needs and goals presented.

XII.10   Special Considerations for Online and Telephone Services

XII.10.01 A service agreement is established between the client and staff member or affiliate before telephone and/or online modalities are used to provide EAP services.

XII.10.02 The EAP uses telephone and/or online services in one or more of the following circumstances:

  a. pursuant to a contract with the customer;
  b. the client is remote and thus suitable for telephone or online services;
  c. the client has mobility problems; or
  d. a supervisor determines there are other extenuating circumstances that merit the use of telephone or online services.

XII.10.03 The EAP requires all telephone counselors and affiliates to use a hard-wired, non-portable telephone and to inform the client of optimal circumstances for the delivery of telephone services, such as:

    a. allocation of sufficient time free from non-emergent interruptions; and

    b. the non-use of cordless or cellular telephones, except in an emergency.

Interpretation (XII.10.03):
Portable telephones are permissible in emergency situations.

XII.11   International

XII.11.01 EAPs that serve their clients internationally ensure that all EAP providers are qualified according to the relevant standards in this section and possess terminal degrees from the countries in which they operate, when applicable.

XII.11.02 EAPs based in a foreign country ensure that all EAP providers are qualified according to the requirements of their respective locations.

Note: EAP companies based in one country, but providing services in another country, must demonstrate that their identified network of resources is active, meets local licensure and/or certification requirements, and is eligible for liability insurance.

XII.11.03 International EAPs adhere to all laws and customs in the countries in which they operate and serve clients.

Interpretation (XII.11.03):
This standard applies to multinational corporations that serve employees abroad, as well as EAPs that are incorporated in a foreign country, and includes all country-specific issues related to confidentiality, record keeping, and legal mandates impacting the workplace and EAP service delivery, such as drug testing.

XII.11.04 International EAPs adhere to a code of ethics created internally or through a professional membership organization that addresses standards related to professional competency and ongoing staff development, conduct, business practices, record-keeping, and confidentiality.

XII.11.05 All international staff and affiliates receive an orientation to the EAP's ethical standards and/or a copy of such standards.

XII.11.06 The program design of international EAPs incorporates mechanisms for communication and promotion of the program in a culturally sensitive fashion.

Interpretation (XII.11.06):
This includes communication materials, supervisor training, and employee orientations, as defined by the standards.

XII.11.07 Where impossible to comply with these standards due to country-specific barriers, EAP providers have a defined policy for follow-up with EAP clients that is culturally sensitive and promotes quality of services delivered, optimal outcome results, and participant satisfaction.

Index

T - #0550 - 101024 - C0 - 212/152/11 - PB - 9780789026446 - Gloss Lamination